Tiu Kiu Wong

THEORY OF H^p SPACES

This is Volume 38 in
PURE AND APPLIED MATHEMATICS
A Series of Monographs and Textbooks
Editors: PAUL A. SMITH AND SAMUEL EILENBERG
A complete list of titles in this series appears at the end of this volume

THEORY OF H^p SPACES

Peter L. Duren
Department of Mathematics
University of Michigan
Ann Arbor, Michigan

Academic Press

New York and London
1970

ACADEMIC PRESS, INC.
111 Fifth Avenue, New York, New York 10003

United Kingdom Edition published by
ACADEMIC PRESS, INC. (LONDON) LTD.
Berkeley Square House, London W1X 6BA

LIBRARY OF CONGRESS CATALOG CARD NUMBER: 74-117092

PRINTED IN THE UNITED STATES OF AMERICA

TO MY FATHER

William L. Duren

CONTENTS

PREFACE

The theory of H^p spaces has its origins in discoveries made forty or fifty years ago by such mathematicians as G. H. Hardy, J. E. Littlewood, I. I. Privalov, F. and M. Riesz, V. Smirnov, and G. Szegö. Most of this early work is concerned with the properties of individual functions of class H^p, and is classical in spirit. In recent years, the development of functional analysis has stimulated new interest in the H^p classes as linear spaces. This point of view has suggested a variety of natural problems and has provided new methods of attack, leading to important advances in the theory.

This book is an account of both aspects of the subject, the classical and the modern. It is intended to provide a convenient source for the older parts of the theory (the work of Hardy and Littlewood, for example), as well as to give a self-contained exposition of more recent developments such as Beurling's theorem on invariant subspaces, the Macintyre–Rogosinski–Shapiro–Havinson theory of extremal problems, interpolation theory, the dual space structure of H^p with $p < 1$, H^p spaces over general domains, and Carleson's proof of the corona theorem. Some of the older results are proved by modern methods. In fact, the dominant theme of the book is the interplay of "hard" and "soft" analysis, the blending of classical and modern techniques and viewpoints.

The book should prove useful both to the research worker and to the graduate student or mathematician who is approaching the subject for the first time. The only prerequisites are an elementary working knowledge of real and complex analysis, including Lebesgue integration and the elements of functional analysis. For example, the books (cited in the bibliography) of Ahlfors or Titchmarsh, Natanson or Royden, and Goffman and Pedrick are more than adequate background. Occasionally, particularly in the last few chapters, some more advanced results enter into the discussion, and appropriate references are given. But the book is essentially self-contained, and it can serve as a textbook for a course at the second- or third-year graduate level. In fact, the book has evolved from lectures which I gave in such a course at the University of Michigan in 1964 and again in 1966. With the student in mind, I have tried to keep things at an elementary level wherever possible.

On the other hand, some sections of the book (for example, parts of Chapters 4, 6, 7, 9, 10, and 12) are rather specialized and are directed primarily to research workers. Many of these topics appear for the first time in book form. In particular, the last chapter, which gives a complete proof of the corona theorem, is "for adults only."

Each chapter contains a list of exercises. Some of them are straightforward, others are more challenging, and a few are quite difficult. Those in the last category are usually accompanied by references to the literature. Many of the exercises point out directions in which the theory can be extended and applied. Further indications of this type, as well as historical remarks and references, appear in the *Notes* at the end of each chapter. Two appendices are included to develop background material which the average mathematician cannot be expected to know.

The chapters need not be read in sequence. For example, Chapters 8 and 9 depend only upon the first three chapters (with some deletions possible) and upon the first two sections of Chapter 7. Chapter 12 can be read immediately after Chapters 8 and 9.

The coverage is reasonably complete, but some topics which might have been included are mentioned only in the *Notes*, or not at all. Inevitably, my own interests have influenced the selection of material.

I wish to express my sincere appreciation to the many friends, students, and colleagues who offered valuable advice or criticized earlier versions of the manuscript. I am especially indebted to J. Caughran, W. L. Duren, F. W. Gehring, W. K. Hayman, J. Hesse, H. J. Landau, A. Macdonald, B. Muckenhoupt, P. Rosenthal, W. Rudin, J. V. Ryff, D. Sarason, H. S. Shapiro, A. L. Shields, B. A. Taylor, G. D. Taylor, G. Weiss and A. Zygmund. I am very thankful to my wife Gay, who accurately prepared the bibliography and proofread the entire book. Renate McLaughlin's help with the proofreading was also most valuable.

In addition, I am grateful to the Alfred P. Sloan Foundation for support during the academic year 1964–1965, when I wrote the first coherent draft of the book. I had the good fortune to spend this year at Imperial College, University of London and at the Centre d'Orsay, Université de Paris. The scope of the book was broadened as a result of my mathematical experiences at both of these institutions. In 1968–1969, while at the Institute for Advanced Study on sabbatical leave from the University of Michigan, I added major sections and made final revisions. I am grateful to the National Science Foundation for partial support during this period.

Peter L. Duren

THEORY OF H^p SPACES

This chapter begins with the classical representation theorems for certain classes of harmonic functions in the unit disk, together with some basic results on boundary behavior. After this comes a brief discussion of sub-harmonic functions. Both topics are fundamental to the theory of H^p spaces. In particular, subharmonic functions provide a strikingly simple approach to Hardy's convexity theorem and to Littlewood's subordination theorem, as shown in Sections 1.4 and 1.5. Finally, the Hardy–Littlewood maximal theorem (proved in Appendix B) is applied to establish an important maximal theorem for analytic functions.

1.1. HARMONIC FUNCTIONS

Many problems of analysis center upon analytic functions with restricted growth near the boundary. For functions analytic in a disk, the integral means

$$M_p(r,f) = \left\{ \frac{1}{2\pi} \int_0^{2\pi} |f(re^{i\theta})|^p \, d\theta \right\}^{1/p}, \qquad 0 < p < \infty;$$

$$M_\infty(r,f) = \max_{0 \le \theta < 2\pi} |f(re^{i\theta})|$$

provide one measure of growth and lead to a particularly rich theory with broad applications. A function $f(z)$ analytic in the unit disk $|z| < 1$ is said to be of class H^p $(0 < p \leq \infty)$ if $M_p(r, f)$ remains bounded as $r \to 1$. Thus H^∞ is the class of bounded analytic functions in the disk, while H^2 is the class of power series $\sum a_n z^n$ with $\sum |a_n|^2 < \infty$.

It is convenient also to introduce the analogous classes of harmonic functions. A real-valued function $u(z)$ harmonic in $|z| < 1$ is said to be of class h^p $(0 < p \leq \infty)$ if $M_p(r, u)$ is bounded. Since

$$a^p \leq (a + b)^p \leq 2^p(a^p + b^p), \qquad a \geq 0, \quad b \geq 0,$$

for $0 < p < \infty$, an analytic function belongs to H^p if and only if its real and imaginary parts are both in h^p. The same inequality shows that H^p and h^p are linear spaces. Finally, it is evident that $H^p \supset H^q$ if $0 < p < q \leq \infty$, and likewise for the h^p spaces.

Any real-valued function $u(z)$ harmonic in $|z| < 1$ and continuous in $|z| \leq 1$ can be recovered from its boundary function by the *Poisson integral*

$$u(z) = u(re^{i\theta}) = \frac{1}{2\pi} \int_0^{2\pi} P(r, \theta - t) u(e^{it}) \, dt, \tag{1}$$

where

$$P(r, \theta) = \frac{1 - r^2}{1 - 2r \cos \theta + r^2}$$

is the *Poisson kernel*. Now replace $u(e^{it})$ in the integral (1) by an arbitrary continuous function $\varphi(t)$ with $\varphi(0) = \varphi(2\pi)$. The resulting function $u(z)$ is still harmonic in $|z| < 1$, continuous in $|z| \leq 1$, and has boundary values $u(e^{it}) = \varphi(t)$. Generalizing this idea, one is led to the notion of a *Poisson–Stieltjes integral*. This is a function of the form

$$u(z) = u(re^{i\theta}) = \frac{1}{2\pi} \int_0^{2\pi} P(r, \theta - t) \, d\mu(t), \tag{2}$$

where $\mu(t)$ is of bounded variation on $[0, 2\pi]$. Again, each such function is harmonic in $|z| < 1$.

THEOREM 1.1. The following three classes of functions in $|z| < 1$ are identical:
(i) Poisson–Stieltjes integrals;
(ii) differences of two positive harmonic functions;
(iii) h^1.

The proof is based on the Helly selection theorem, which we now state for the convenience of the reader. (For a proof, see Natanson [1] or Widder [1]. Also, see *Notes*.)

LEMMA (Helly selection theorem). Let $\{\mu_n(t)\}$ be a uniformly bounded sequence of functions of uniformly bounded variation over a finite interval $[a, b]$. Then some subsequence $\{\mu_{n_k}(t)\}$ converges everywhere in $[a, b]$ to a function $\mu(t)$ of bounded variation, and for every continuous function $\varphi(t)$,

$$\lim_{k \to \infty} \int_a^b \varphi(t) \, d\mu_{n_k}(t) = \int_a^b \varphi(t) \, d\mu(t).$$

PROOF OF THEOREM 1.1. (i) $\Rightarrow$ (ii). Expressing $\mu(t)$ as the difference of two bounded nondecreasing functions, we see that every Poisson–Stieltjes integral is the difference of two positive harmonic functions.

(ii) $\Rightarrow$ (iii). Suppose $u(z) = u_1(z) - u_2(z)$, where u_1 and u_2 are positive harmonic functions. Then

$$\int_0^{2\pi} |u(re^{i\theta})| \, d\theta \le \int_0^{2\pi} u_1(re^{i\theta}) \, d\theta + \int_0^{2\pi} u_2(re^{i\theta}) \, d\theta$$

$$= 2\pi[u_1(0) + u_2(0)],$$

so that $u \in h^1$.

(iii) $\Rightarrow$ (i). Given $u \in h^1$, define

$$\mu_r(t) = \int_0^t u(re^{i\theta}) \, d\theta.$$

Then $\mu_r(0) = 0$, and for $0 = t_0 < t_1 < \cdots < t_n = 2\pi$,

$$\sum_{k=1}^n |\mu_r(t_k) - \mu_r(t_{k-1})| \le \int_0^{2\pi} |u(re^{i\theta})| \, d\theta \le C.$$

Hence the functions $\mu_r(t)$ are of uniformly bounded variation. By the Helly selection theorem, there is a sequence $\{r_n\}$ tending to 1 for which $\mu_{r_n}(t) \to \mu(t)$, a function of bounded variation in $0 \le t \le 2\pi$. Thus

$$\frac{1}{2\pi} \int_0^{2\pi} P(r, \theta - t) \, d\mu(t) = \lim_{n \to \infty} \frac{1}{2\pi} \int_0^{2\pi} P(r, \theta - t) \, d\mu_{r_n}(t)$$

$$= \lim_{n \to \infty} \frac{1}{2\pi} \int_0^{2\pi} P(r, \theta - t) u(r_n e^{it}) \, dt = \lim_{n \to \infty} u(r_n z) = u(z).$$

(Here, as always, $z = re^{i\theta}$.)

As a corollary to the proof, we see that every positive harmonic function in the unit disk can be represented as a Poisson–Stieltjes integral with respect to a *nondecreasing* function $\mu(t)$. This is usually called the *Herglotz representation*. The function $\mu(t)$ of bounded variation corresponding to a given $u \in h^1$ is essentially unique. Indeed, if $\int P(r, \theta - t)\, d\mu(t) \equiv 0$, analytic completion gives

$$\int_0^{2\pi} \frac{e^{it} + z}{e^{it} - z}\, d\mu(t) \equiv i\gamma, \qquad |z| < 1, \quad z = re^{i\theta},$$

where γ is a real constant. Since

$$\frac{e^{it} + z}{e^{it} - z} = 1 + 2\sum_{n=1}^{\infty} e^{-int} z^n,$$

we conclude that

$$\int_0^{2\pi} e^{int}\, d\mu(t) = 0, \qquad n = 0, \pm 1, \pm 2, \dots$$

Since the characteristic function of any interval can be approximated in L^1 by a continuous periodic function, hence by a trigonometric polynomial, this shows that the measure of each interval is zero. Thus $d\mu$ is the zero measure.

1.2. BOUNDARY BEHAVIOR OF POISSON–STIELTJES INTEGRALS

If $u(z)$ is the Poisson integral of an integrable function $\varphi(t)$, then for any point $t = \theta_0$ where φ is continuous, $u(z) \to \varphi(\theta_0)$ as $z \to e^{i\theta_0}$. This can be generalized to Poisson-Stieltjes integrals: $u(z) \to \mu'(\theta_0)$ wherever μ is continuously differentiable. Actually, it is enough that μ be differentiable; or, slightly more generally, that the *symmetric derivative*

$$D\mu(\theta_0) = \lim_{t \to 0} \frac{\mu(\theta_0 + t) - \mu(\theta_0 - t)}{2t}$$

exist, as the following theorem shows.

Fatou

THEOREM 1.2. Let $u(z)$ be a Poisson–Stieltjes integral of the form (2), where μ is of bounded variation. If the symmetric derivative $D\mu(\theta_0)$ exists at a point θ_0, then the radial limit $\lim_{r \to 1} u(re^{i\theta_0})$ exists and has the value $D\mu(\theta_0)$.

PROOF. We may assume $\theta_0 = 0$. Set $A = D\mu(0)$, and write

$$u(r) - A = \frac{1}{2\pi} \int_{-\pi}^{\pi} P(r, t)[d\mu(t) - A\, dt]$$

$$= \frac{1}{2\pi} [P(r, t)[\mu(t) - At]]_{-\pi}^{\pi}$$

$$- \frac{1}{2\pi} \int_{-\pi}^{\pi} [\mu(t) - At]\left[\frac{\partial}{\partial t} P(r, t)\right] dt.$$

The integrated term tends to zero as $r \to 1$. For $0 < \delta \le |t| \le \pi$,

$$\left| \frac{\partial P}{\partial t} \right| \le \frac{2r(1 - r^2)}{[1 - 2r \cos \delta + r^2]^2} \to 0 \qquad \text{as} \quad r \to 1.$$

Hence for each fixed $\delta > 0$, $u(r) - A - I_\delta \to 0$, where

$$I_\delta = -\frac{1}{2\pi} \int_{-\delta}^{\delta} [\mu(t) - At] \left[\frac{\partial}{\partial t} P(r, t) \right] dt$$

$$= \frac{1}{\pi} \int_0^{\delta} \left[\frac{\mu(t) - \mu(-t)}{2t} - A \right] t \left[-\frac{\partial}{\partial t} P(r, t) \right] dt.$$

Given $\varepsilon > 0$, choose $\delta > 0$ so small that

$$\left| \frac{\mu(t) - \mu(-t)}{2t} - A \right| \le \varepsilon \qquad \text{for} \quad 0 < t \le \delta.$$

Then

$$|I_\delta| \le \frac{\varepsilon}{2\pi} \int_{-\pi}^{\pi} t \left(-\frac{\partial P}{\partial t} \right) dt < 2\varepsilon$$

for r sufficiently near 1, as an integration by parts shows. Thus $u(r) \to A$ as $r \to 1$, and the proof is complete.

Since a function of bounded variation is differentiable almost everywhere, we obtain two important corollaries.

COROLLARY 1. Each function $u \in h^1$ has a radial limit almost everywhere.

COROLLARY 2. If u is the Poisson integral of a function $\varphi \in L^1$, then $u(re^{i\theta}) \to \varphi(\theta)$ almost everywhere.

By a refinement of the proof it is even possible to show that $u(z)$ tends to $D\mu(\theta_0)$ along any path not tangent to the unit circle. However, we shall arrive at this result (almost everywhere) by an indirect route. For the present, we content ourselves with showing that a bounded analytic function has such a nontangential limit almost everywhere.

For $0 < \alpha < \pi/2$, construct the sector with vertex $e^{i\theta}$, of angle 2α, symmetric with respect to the ray from the origin through $e^{i\theta}$. Draw the two segments from the origin perpendicular to the boundaries of this sector, and let $S_\alpha(\theta)$ denote the "kite-shaped" region so constructed (see Fig. 1).

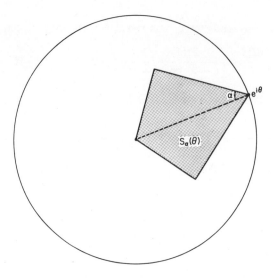

Figure 1

THEOREM 1.3. If $f \in H^\infty$, the radial limit $\lim_{r \to 1} f(re^{i\theta})$ exists almost everywhere. Furthermore, if θ_0 is a value for which the radial limit exists, then $f(z)$ tends to the same limit as $z \to e^{i\theta_0}$ inside any region $S_\alpha(\theta_0)$, $\alpha < \pi/2$.

PROOF. The existence almost everywhere of a radial limit follows from Corollary 1 to Theorem 1.2, since $h^\infty \subset h^1$. To discuss the angular limit, it is convenient to deal instead with a bounded analytic function $f(z)$ in the disk $|z - 1| < 1$, having a limit L as $z \to 0$ along the positive real axis. Let $f_n(z) = f(z/n)$, $n = 1, 2, \ldots$. The functions $f_n(z)$ are uniformly bounded, so they form a *normal family* (see Ahlfors [2], Chap. 5). This implies that a subsequence tends to an analytic function $F(z)$ uniformly in each closed subdomain of the disk, hence in the region

$$|\arg z| \le \alpha < \pi/2; \qquad (\cos \alpha)/2 \le |z| \le \cos \alpha. \tag{3}$$

(The ray $\arg z = \alpha$ has a segment of length $2 \cos \alpha$ in common with the disk $|z - 1| < 1$.) But for all real z in the interval $0 < z < 2$, $f_n(z) \to L$. It follows that $F(z) \equiv L$, and that $f_n(z) \to L$ uniformly in the region (3). This implies that $f(z) \to L$ as $z \to 0$ inside the sector $|\arg z| \le \alpha$, which proves the theorem.

The function f is said to have a *nontangential limit* L at $e^{i\theta_0}$ if $f(z) \to L$ as $z \to e^{i\theta_0}$ inside each region $S_\alpha(\theta_0)$, $\alpha < \pi/2$. Thus each $f \in H^\infty$ has a nontangential limit almost everywhere.

1.3. SUBHARMONIC FUNCTIONS

A *domain* is an open connected set in the complex plane. Let D be a bounded domain. The *boundary* ∂D of D is defined to be the closure $\bar{D}$ minus D. A real-valued function $g(z)$ continuous in D is said to be *subharmonic* if it has the following property. For each domain B with $\bar{B} \subset D$, and for each function $U(z)$ harmonic in B, continuous in $\bar{B}$, such that $g(z) \leq U(z)$ on ∂B, the inequality $g(z) \leq U(z)$ holds throughout B. In particular, if there is a function $U(z)$ harmonic in B with boundary values $g(z)$, then $g(z) \leq U(z)$ in B.

Subharmonic functions are also characterized by the "local sub-mean-value property," which is often easier to work with.

THEOREM 1.4. Necessary and sufficient that a continuous function $g(z)$ be subharmonic in D is that for each $z_0 \in D$ there exist $\rho_0 > 0$ such that the disk $|z - z_0| < \rho_0$ is in D and

$$g(z_0) \leq \frac{1}{2\pi} \int_0^{2\pi} g(z_0 + \rho e^{i\theta})\, d\theta \qquad (4)$$

for every $\rho < \rho_0$.

PROOF. The necessity is easy. Let $|z - z_0| < \rho$ be in D, and let $U(z)$ be the function harmonic in this disk and equal to $g(z)$ on $|z - z_0| = \rho$. Then

$$g(z_0) \leq U(z_0) = \frac{1}{2\pi} \int_0^{2\pi} U(z_0 + \rho e^{i\theta})\, d\theta = \frac{1}{2\pi} \int_0^{2\pi} g(z_0 + \rho e^{i\theta})\, d\theta.$$

To prove the sufficiency, suppose there exists a domain B with $\bar{B} \subset D$ and a harmonic function $U(z)$ such that $g(z) \leq U(z)$ on ∂B, yet $g(z) > U(z)$ somewhere in B. Let E be the set of points in $\bar{B}$ at which $h(z) = g(z) - U(z)$ attains its maximum $m > 0$. Then $E \subset B$, because $h(z) \leq 0$ on ∂B. Since E is a closed set, some point $z_0 \in E$ has no circular neighborhood entirely contained in E. Hence there exists a sequence $\{\rho_n\}$ tending to zero such that the disk $|z - z_0| < \rho_n$ is in B, while the circle $|z - z_0| = \rho_n$ is not entirely contained in E. Thus $h(z) \leq m$ on $|z - z_0| = \rho_n$, with strict inequality on an open subset of the circle, hence on an arc. It follows that

$$\frac{1}{2\pi} \int_0^{2\pi} g(z_0 + \rho_n e^{i\theta})\, d\theta - U(z_0) = \frac{1}{2\pi} \int_0^{2\pi} h(z_0 + \rho_n e^{i\theta})\, d\theta$$

$$< m = h(z_0) = g(z_0) - U(z_0).$$

But this contradicts property (4), so the proof is complete.

EXAMPLE 1. If $f(z)$ is analytic in a domain D and $p > 0$, then $g(z) = |f(z)|^p$ is subharmonic in D. To see this, we need only verify (4). If $f(z_0) = 0$, there is nothing to prove. If $f(z_0) \neq 0$, then some branch of $[f(z)]^p$ is analytic in a neighborhood of z_0, so that

$$[f(z_0)]^p = \frac{1}{2\pi} \int_0^{2\pi} [f(z_0 + \rho e^{i\theta})]^p \, d\theta$$

for sufficiently small ρ. From this (4) easily follows.

EXAMPLE 2. If $u(z)$ is harmonic in D and $p \geq 1$, then $|u(z)|^p$ is subharmonic in D. Again, we have only to check (4). This is obvious if $p = 1$. If $p > 1$, we apply Hölder's inequality to the mean value relation and obtain

$$|u(z_0)| \leq \frac{1}{2\pi} \left[\int_0^{2\pi} |u(z_0 + \rho e^{i\theta})|^p \, d\theta \right]^{1/p} (2\pi)^{1/q},$$

where $1/p + 1/q = 1$. This implies (4).

EXAMPLE 3. Let

$$\log^+ x = \begin{cases} \log x, & x \geq 1 \\ 0, & 0 \leq x < 1. \end{cases}$$

Then $g(z) = \log^+ |f(z)|$ is subharmonic if $f(z)$ is analytic. Indeed, (4) is trivially true if $|f(z_0)| \leq 1$. If $|f(z_0)| > 1$, then this inequality persists in some neighborhood $|z - z_0| \leq \rho$. Thus $\log^+|f(z)|$ coincides with the harmonic function $\log|f(z)|$ in $|z - z_0| \leq \rho$.

1.4. HARDY'S CONVEXITY THEOREM

The class H^p was introduced in Section 1.1 as the set of all functions $f(z)$ analytic in $|z| < 1$ for which the means $M_p(r, f)$ are bounded. It is natural to ask how these means may behave as functions of r, for an arbitrary analytic function f.

The case $p = 2$ is especially simple. If $f(z) = \sum a_n z^n$ is analytic in $|z| < 1$, then by Parseval's relation

$$M_2^2(r, f) = \sum_{n=0}^{\infty} |a_n|^2 r^{2n}.$$

This shows that $M_2(r, f)$ increases with r, and that $f \in H^2$ if and only if $\sum |a_n|^2 < \infty$. Likewise, it follows from the maximum modulus principle that $M_\infty(r, f)$ increases with r.

The situation is more complicated for other values of p, but $M_p(r, f)$ is always a nondecreasing function. In fact, much more is true.

THEOREM 1.5 (Hardy's convexity theorem). Let $f(z)$ be analytic in $|z| < 1$, and let $0 < p \leq \infty$. Then
  (i) $M_p(r, f)$ is a nondecreasing function of r;
  (ii) $\log M_p(r, f)$ is a convex function of $\log r$.

To say that $\log M_p(r, f)$ is a convex function of $\log r$ means that if

$$\log r = \alpha \log r_1 + (1 - \alpha) \log r_2 \qquad (0 < r_1 < r_2 < 1; 0 < \alpha < 1),$$

then

$$\log M_p(r, f) \leq \alpha \log M_p(r_1, f) + (1 - \alpha) \log M_p(r_2, f),$$

or

$$M_p(r, f) \leq [M_p(r_1, f)]^\alpha [M_p(r_2, f)]^{1-\alpha}.$$

For $p = \infty$, this is the well-known Hadamard "three-circles" theorem (see Titchmarsh [1], p. 172), which is an easy consequence of the maximum modulus principle.

We shall deduce Hardy's theorem from a general theorem on subharmonic functions.

THEOREM 1.6. Let $g(z)$ be subharmonic in $|z| < 1$, and let

$$m(r) = \frac{1}{2\pi} \int_0^{2\pi} g(re^{i\theta}) \, d\theta, \qquad 0 \leq r < 1.$$

Then $m(r)$ is nondecreasing and is a convex function of $\log r$.

PROOF. Choose $0 \leq r_1 < r_2 < 1$, and let $U(z)$ be the function harmonic in $|z| < r_2$, continuous in $|z| \leq r_2$, and equal to $g(z)$ on $|z| = r_2$. Then $g(z) \leq U(z)$ in $|z| \leq r_2$, so

$$m(r_1) \leq \frac{1}{2\pi} \int_0^{2\pi} U(r_1 e^{i\theta}) \, d\theta = U(0)$$

$$= \frac{1}{2\pi} \int_0^{2\pi} U(r_2 e^{i\theta}) \, d\theta = m(r_2).$$

To prove the convexity, let $0 < r_1 < r_2 < 1$, and let $U(z)$ be the function harmonic in the annulus $r_1 < |z| < r_2$ and agreeing with $g(z)$ on both boundary circles. Then

$$m(r) \leq \frac{1}{2\pi} \int_0^{2\pi} U(re^{i\theta}) \, d\theta, \qquad r_1 \leq r \leq r_2, \tag{5}$$

with equality at the end points. On the other hand, by Green's theorem,

$$\frac{d}{dr}\left\{\int_0^{2\pi} U(re^{i\theta})\,d\theta\right\} = \frac{1}{r}\int_0^{2\pi}\frac{\partial U}{\partial n}\,ds = \frac{c}{r},$$

where $\partial/\partial n$ denotes the normal derivative, $ds = r\,d\theta$ is the element of arc-length, and c is a constant. The right-hand side of (5) therefore has the form $a\log r + b$, which shows that $m(r)$ is a convex function of $\log r$.

REMARK. The proof shows that $m(r)$ is a convex function of $\log r$ if $g(z)$ is subharmonic only in an annulus $\rho_1 < |z| < \rho_2$.

PROOF OF HARDY'S THEOREM. (i) As pointed out in Section 1.3, $g(z) = |f(z)|^p$ is subharmonic if $f(z)$ is analytic, so Theorem 1.6 applies. In fact, the same is true of $|u(z)|^p$ if $u(z)$ is harmonic and $p \geq 1$, so $M_p(r, u)$ is also non-decreasing.

(ii) The argument of Section 1.3 shows that $g(z) = |z|^\lambda |f(z)|^p$ is subharmonic in $0 < |z| < 1$ if $p > 0$, λ is any real number, and $f(z)$ is analytic. Thus, by the remark following Theorem 1.6, $r^\lambda M_p{}^p(r, f)$ is a convex function of $\log r$. Given $0 < r_1 < r_2 < 1$, let $\lambda < 0$ be chosen so that

$$r_1{}^\lambda M_p{}^p(r_1, f) = r_2{}^\lambda M_p{}^p(r_2, f) = K,$$

say. Let $r = r_1{}^\alpha r_2{}^{1-\alpha}$ ($0 < \alpha < 1$). Then

$$r^\lambda M_p{}^p(r, f) \leq K = K^\alpha K^{1-\alpha} = \{r_1{}^\lambda M_p{}^p(r_1, f)\}^\alpha \{r_2{}^\lambda M_p{}^p(r_2, f)\}^{1-\alpha}$$
$$= r^\lambda \{M_p{}^p(r_1, f)\}^\alpha \{M_p{}^p(r_2, f)\}^{1-\alpha},$$

which completes the proof.

1.5. SUBORDINATION

Let $F(z)$ be analytic and univalent in $|z| < 1$, with $F(0) = 0$. Let $f(z)$ be analytic in $|z| < 1$, with $f(0) = 0$, and suppose the range of f is contained in that of F. Then $\omega(z) = F^{-1}(f(z))$ is well-defined and analytic in $|z| < 1$, $\omega(0) = 0$, and $|\omega(z)| \leq 1$. By Schwarz's lemma, then, $|\omega(z)| \leq |z|$. This implies, in particular, that the image under $f(z) = F(\omega(z))$ of each disk $|z| \leq r < 1$ is contained in the image of the same disk under $F(z)$.

More generally, a function $f(z)$ analytic in $|z| < 1$ is said to be *subordinate* to an analytic function $F(z)$ (written $f \prec F$) if $f(z) = F(\omega(z))$ for some function $\omega(z)$ analytic in $|z| < 1$, satisfying $|\omega(z)| \leq |z|$. $F(z)$ need not be univalent.

The following result has many applications.

THEOREM 1.7 (Littlewood's subordination theorem). Let $f(z)$ and $F(z)$ be analytic in $|z| < 1$, and suppose $f \prec F$. Then $M_p(r, f) \leq M_p(r, F)$, $0 < p \leq \infty$.

PROOF. As in the proof of Hardy's theorem, we shall deduce this from a more general result concerning subharmonic functions. Let $G(z)$ be subharmonic in $|z| < 1$, and let $g(z) = G(\omega(z))$, where $\omega(z)$ is analytic in $|z| < 1$ and $|\omega(z)| \leq |z|$. Then

$$\int_0^{2\pi} g(re^{i\theta})\, d\theta \leq \int_0^{2\pi} G(re^{i\theta})\, d\theta. \tag{6}$$

To prove (6), from which the theorem easily follows, let $U(z)$ be the function harmonic in $|z| < r$ and equal to $G(z)$ on $|z| = r$. Then $G(z) \leq U(z)$ in $|z| \leq r$, so that $g(z) \leq u(z) = U(\omega(z))$ on $|z| = r$. Hence

$$\frac{1}{2\pi} \int_0^{2\pi} g(re^{i\theta})\, d\theta \leq \frac{1}{2\pi} \int_0^{2\pi} u(re^{i\theta})\, d\theta = u(0)$$

$$= U(0) = \frac{1}{2\pi} \int_0^{2\pi} U(re^{i\theta})\, d\theta = \frac{1}{2\pi} \int_0^{2\pi} G(re^{i\theta})\, d\theta,$$

which proves (6).

1.6. MAXIMAL THEOREMS

For $\varphi \in L^p = L^p(0, 2\pi)$, let

$$\|\varphi\| = \|\varphi\|_p = \left\{ \frac{1}{2\pi} \int_0^{2\pi} |\varphi(t)|^p\, dt \right\}^{1/p}.$$

If $u(r, \theta)$ is the Poisson integral of a function $\varphi \in L^p$, $1 \leq p \leq \infty$, an easy calculation based on Hölder's inequality shows that $u \in h^p$ and $M_p(r, u) \leq \|\varphi\|_p$. There is, however, a much deeper theorem which extends this result in a striking and useful way.

THEOREM 1.8 (Hardy–Littlewood). Let $u(r, \theta)$ be the Poisson integral of $\varphi \in L^p$, $1 < p \leq \infty$, and let

$$U(\theta) = \sup_{r < 1} |u(r, \theta)|.$$

Then $U \in L^p$, and there is a constant A_p depending only on p such that

$$\|U\|_p \leq A_p \|\varphi\|_p.$$

PROOF. The proof depends upon a theorem on "maximal rearrangements" of functions, which is discussed in Appendix B. Extend φ periodically with period 2π, and let

$$\mathcal{M}(\theta) = \mathcal{M}(\theta; \varphi) = \sup_{0 < |T| \leq \pi} \left| \frac{1}{T} \int_0^T \varphi(\theta + t)\, dt \right|.$$

By Theorem B.3, $\mathcal{M} \in L^p$ and

$$\|\mathcal{M}\|_p \le C_p \|\varphi\|_p, \qquad 1 < p \le \infty.$$

$U(\theta)$ is a measurable function, since it is the upper envelope of the functions $|u(r, \theta)|$ as r runs through the sequence of rational numbers in the interval $(0, 1)$. Hence it will suffice to show that

$$|u(r, \theta)| \le 2\mathcal{M}(\theta), \qquad 0 \le r < 1. \tag{7}$$

This can be seen by writing the Poisson formula in the form

$$u(r, \theta) = \frac{1}{2\pi} \int_{-\pi}^{\pi} P(r, t)\varphi(\theta + t)\, dt$$

and integrating by parts. For fixed θ, let

$$\Phi(t) = \int_0^t \varphi(\theta + u)\, du.$$

Then

$$u(r, \theta) = \frac{1}{2\pi} [P(r, t)\Phi(t)]_{-\pi}^{\pi} - \frac{1}{2\pi} \int_{-\pi}^{\pi} t\frac{\partial}{\partial t} \{P(r, t)\} \frac{\Phi(t)}{t}\, dt.$$

From this (7) follows, since $|\Phi(t)/t| \le \mathcal{M}(\theta)$ and

$$\frac{1}{2\pi} \int_{-\pi}^{\pi} \left| t\frac{\partial}{\partial t} P(r, t) \right| dt = \frac{2r}{1 + r} < 1.$$

It should be observed that (7) remains true in the case $p = 1$, although the theorem is then false. (See Exercise 6.) However, the corresponding theorem for *analytic* functions is true for all p in the range $0 < p \le \infty$.

THEOREM 1.9 (Hardy–Littlewood). Let $f \in H^p$, $0 < p \le \infty$, and let

$$F(\theta) = \sup_{r < 1} |f(re^{i\theta})|.$$

Then $F \in L^p$ and

$$\|F\|_p \le B_p \|f\|_p,$$

where B_p depends only on p and

$$\|f\|_p = \lim_{r \to 1} M_p(r, f).$$

PROOF. For each fixed R, $0 < R < 1$, the function $g(z) = |f(Rz)|^{p/2}$ is subharmonic in $|z| < 1$, so it follows from Theorem 1.8 that

$$G(\theta) = \sup_{r < 1} |g(re^{i\theta})| \in L^2$$

and
$$\|G\|_2 \le A_2 M_2(1, g).$$

In other words,
$$\|F_R\|_p \le A_2^{2/p} M_p(R, f),$$

where
$$F_R(\theta) = \sup_{r < R} |f(re^{i\theta})|.$$

Since $F_R(\theta)$ increases to $F(\theta)$ as $R \to 1$, the proof is completed by appeal to the Lebesgue monotone convergence theorem.

EXERCISES

1. Show that $(1 - z)^{-1}$ is in H^p for every $p < 1$, but is not in H^1.

2. Show that if $f(z)$ is analytic in $|z| < 1$ and its range is contained in a sector of angle α $(0 < \alpha \le 2\pi)$, then $f \in H^p$ for every $p < \pi/\alpha$.

3. Show that the function
$$\frac{1}{1 - z} \left(\frac{1}{z} \log \frac{1}{1 - z} \right)^{-c}$$

belongs to H^1 if $c > 1$.

4. Show directly (i.e., by refining the proof of Theorem 1.2) that a Poisson–Stieltjes integral has a nontangential limit almost everywhere.

5. Use Littlewood's subordination theorem to show that $M_p(r, f)$ is a nondecreasing function of r.

6. Use the modified Poisson kernel
$$u(r, \theta) = \frac{R^2 - r^2}{R^2 - 2Rr \cos \theta + r^2}, \qquad R > 1,$$

to show that Theorem 1.8 is false for $p = 1$.

7. Show that if $f(z)$ and $g(z)$ are subharmonic, then so is $\max\{f(z), g(z)\}$.

NOTES

In the language of functional analysis, the Helly selection theorem simply asserts the weak-star sequential compactness of the unit sphere of the dual space of the space C of continuous functions over the interval $[a, b]$. That is, if $\phi_n \in C^*$ and $\sup \|\phi_n\| < \infty$, then some subsequence $\{\phi_{n_k}\}$ converges pointwise to $\phi \in C^*$.

The paper of Fatou [1] is one of the earliest studies of the boundary behavior of analytic and harmonic functions in the disk. Fatou showed, among other things, that a Poisson–Stieltjes integral has a nontangential limit almost everywhere. The existence of such a limit for a bounded analytic function is known as "Fatou's theorem." Theorem 1.1 goes back at least to Plessner [1], although Herglotz [1] had earlier found the representation for positive harmonic functions. Hardy's convexity theorem is in his paper [2], which is now regarded as the historical starting point of the theory of H^p spaces. (Thus the "H" in "H^p" is for Hardy.) The elegant proof of Hardy's theorem via subharmonic functions is due to F. Riesz [2]. A. E. Taylor [1] found another proof using Banach space methods. Littlewood's subordination theorem is in his paper [1]; again F. Riesz [5] vastly simplified the proof using subharmonic functions. For an extension, see Gabriel [1,2]. See also Ryff [1]. For a thorough discussion of subharmonic functions, see Rado [1]. The Hardy–Littlewood maximal theorems are in their paper [4]. They also proved more general theorems in which the supremum is taken not over a radial segment, but over a "kite-shaped region" $S_\alpha(\theta)$ with vertex at a boundary point.

In this chapter we consider, among other things, the boundary behavior and the zeros of H^p functions. As may be expected, the growth restriction on an H^p function imposes a restriction on the density of its zeros. This leads to the formation of "Blaschke products" and the canonical factorization theorems, which play a vital role in later work. Another important result is the mean convergence of an H^p function to its boundary function. Two new classes of functions, N and N^+, arise in the course of the discussion, and their properties are explored. Finally, an equivalent definition of H^p is given in terms of harmonic majorants.

2.1. BOUNDARY VALUES

We have seen that a harmonic function of class h^1 must have a radial limit in almost every direction. The same is therefore true for analytic functions of class H^1. The theorem remains valid for H^p functions with $p < 1$, despite the fact (to be noted in Chapter 4) that the corresponding statement for harmonic functions is *false*. Surprisingly, the generalization to $p < 1$ is obtained most simply by dealing with an even wider class of functions. This is the *Nevanlinna*

class N, also known as the class of functions of *bounded characteristic.*
A function $f(z)$ analytic in $|z| < 1$ is said to be of class N if the integrals

$$\int_0^{2\pi} \log^+ |f(re^{i\theta})| \, d\theta$$

are bounded for $r < 1$. Since $\log^+ |f|$ is subharmonic whenever f is analytic,
these integrals always increase with r. It is clear that N contains H^p for every
$p > 0$.

The following theorem is basic to the further development of the theory.

THEOREM 2.1 (F. and R. Nevanlinna). A function analytic in the unit
disk belongs to the class N if and only if it is the quotient of two bounded
analytic functions.

PROOF. Suppose first that $f(z) = \varphi(z)/\psi(z)$, where φ and ψ are analytic
and bounded in $|z| < 1$. There is no loss of generality in assuming $|\varphi(z)| \leq 1$,
$|\psi(z)| \leq 1$, and $\psi(0) \neq 0$. Then

$$\int_0^{2\pi} \log^+ |f(re^{i\theta})| \, d\theta \leq - \int_0^{2\pi} \log |\psi(re^{i\theta})| \, d\theta.$$

But by Jensen's formula (see Ahlfors [2], p. 206),

$$\frac{1}{2\pi} \int_0^{2\pi} \log |\psi(re^{i\theta})| \, d\theta = \log |\psi(0)| + \sum_{|z_n| < r} \log \frac{r}{|z_n|},$$

where z_n are the zeros of ψ. This shows that $\int \log |\psi|$ increases with r, so $f \in N$.

Conversely, let $f(z) \not\equiv 0$ be of class N. Let f have a zero of multiplicity
$m \geq 0$ at the origin, so that $z^{-m} f(z) \to \alpha \neq 0$ as $z \to 0$. Let z_n be the other zeros
of f, repeated according to multiplicity and arranged so that $0 < |z_1| \leq$
$|z_2| \leq \cdots < 1$. If $f(z) \neq 0$ on the circle $|z| = \rho < 1$, the function

$$F(z) = \log \left\{ f(z) \frac{\rho^m}{z^m} \prod_{|z_n| < \rho} \left(\frac{\rho^2 - \bar{z}_n z}{\rho(z - z_n)} \right) \right\}$$

is analytic in $|z| \leq \rho$, and $\operatorname{Re}\{F(z)\} = \log |f(z)|$ on $|z| = \rho$. Hence, by analytic
completion of the Poisson formula,

$$F(z) = \frac{1}{2\pi} \int_0^{2\pi} \log |f(\rho e^{it})| \frac{\rho e^{it} + z}{\rho e^{it} - z} \, dt + iC.$$

This is sometimes called the *Poisson–Jensen formula.* After exponentiation,
it takes the form $f(z) = \varphi_\rho(z)/\psi_\rho(z)$, where

$$\varphi_\rho(z) = \frac{z^m}{\rho^m} \prod_{|z_n| < \rho} \frac{\rho(z - z_n)}{\rho^2 - \bar{z}_n z} \cdot \exp\left\{ -\frac{1}{2\pi} \int_0^{2\pi} \log^- |f(\rho e^{it})| \frac{\rho e^{it} + z}{\rho e^{it} - z} \, dt + iC \right\},$$

$$\psi_\rho(z) = \exp\left\{ -\frac{1}{2\pi} \int_0^{2\pi} \log^+ |f(\rho e^{it})| \frac{\rho e^{it} + z}{\rho e^{it} - z} \, dt \right\}.$$

Now choose a sequence $\{\rho_k\}$ increasing to 1, such that $f(z) \neq 0$ on the circles $|z| = \rho_k$. Let

$$\Phi_k(z) = \varphi_{\rho_k}(\rho_k z); \qquad \Psi_k(z) = \psi_{\rho_k}(\rho_k z).$$

Then $f(\rho_k z) = \Phi_k(z)/\Psi_k(z)$ for $|z| < 1$. But the functions Φ_k and Ψ_k are analytic in the unit disk, and $|\Phi_k(z)| \leq 1$, $|\Psi_k(z)| \leq 1$ there. Hence $\{\Phi_k\}$ and $\{\Psi_k\}$ are normal families, and there is a sequence $\{k_i\}$ such that $\Phi_{k_i}(z) \to \varphi(z)$ and $\Psi_{k_i}(z) \to \psi(z)$ uniformly in each disk $|z| \leq R < 1$. The functions φ and ψ are analytic in the unit disk, and $|\varphi(z)| \leq 1$, $|\psi(z)| \leq 1$. According to the definition of ψ_ρ, the fact that $\int \log^+ |f|$ is bounded gives a uniform estimate $|\Psi_k(0)| \geq \delta > 0$, so $\psi(z) \not\equiv 0$. Thus $f = \varphi/\psi$, and the proof is complete.

The importance of this representation theorem is that it allows properties of functions in N to be deduced from the corresponding properties of bounded analytic functions. The boundary behavior, for example, can now be discussed.

THEOREM 2.2. For each function $f \in N$, the nontangential limit $f(e^{i\theta})$ exists almost everywhere, and $\log |f(e^{i\theta})|$ is integrable unless $f(z) \equiv 0$. If $f \in H^p$ for some $p > 0$, then $f(e^{i\theta}) \in L^p$.

PROOF. Assuming $f(z) \not\equiv 0$, represent it in the form $\varphi(z)/\psi(z)$, where $|\varphi(z)| \leq 1$ and $|\psi(z)| \leq 1$. Since φ and ψ are bounded analytic functions, they have nontangential limits $\varphi(e^{i\theta})$ and $\psi(e^{i\theta})$ almost everywhere, by Theorem 1.3. Appealing to Fatou's lemma, we find

$$\int_0^{2\pi} \left| \log |\varphi(e^{i\theta})| \right| \, d\theta \leq \liminf_{r \to 1} \left\{ -\int_0^{2\pi} \log |\varphi(re^{i\theta})| \, d\theta \right\}.$$

But $\int \log |\varphi(re^{i\theta})| \, d\theta$ increases with r, by Jensen's theorem. Hence $\log |\varphi(e^{i\theta})| \in L^1$, and similarly for ψ. In particular, $\psi(e^{i\theta})$ cannot vanish on a set of positive measure. The radial limit $f(e^{i\theta})$ therefore exists almost everywhere, and $\log |f(e^{i\theta})| \in L^1$. Another application of Fatou's lemma shows that $f(e^{i\theta}) \in L^p$ if $f \in H^p$.

The theorem says that if $f \in N$ and if $f(e^{i\theta}) = 0$ on a set of positive measure, then $f(z) \equiv 0$. In other words, a function of class N is uniquely determined by its boundary values on any set of positive measure.

It is evident from the representation $f = \varphi/\psi$ that $\int \log^- |f(re^{i\theta})|\, d\theta$ is bounded if $f \in N$. (Here $\log^- x = \max\{-\log x, 0\}$.) Hence $f \in N$ if and only if $\int \big|\log |f(re^{i\theta})|\big|\, d\theta$ is bounded.

2.2. ZEROS

The zeros of an analytic function cannot cluster inside the domain of analyticity unless the function vanishes identically. More is true if the function satisfies a growth restriction: the zeros must tend "rapidly" to the boundary. The following theorem illustrates this principle.

THEOREM 2.3. Let $f(z) \not\equiv 0$ be analytic in $|z| < 1$, and let $z_1, z_2, \ldots$ be its zeros, repeated according to multiplicity, in $|z| < 1$. Then $\int_0^{2\pi} \log|f(re^{i\theta})|\, d\theta$ is bounded if and only if $\sum_{n=1}^{\infty} (1 - |z_n|) < \infty$.

PROOF. Let f have a zero of multiplicity $m \geq 0$ at the origin, so that $f(z) = \alpha z^m + \cdots$, $\alpha \neq 0$. Call the other zeros a_n, and let them be ordered so that $0 < |a_1| \leq |a_2| \leq \cdots$. By Jensen's formula,

$$\frac{1}{2\pi} \int_0^{2\pi} \log|f(re^{i\theta})|\, d\theta = \sum_{|a_n|<r} \log \frac{r}{|a_n|} + \log(|\alpha| r^m).$$

This increases monotonically with r; suppose it is bounded from above by some constant C. Then, fixing an integer N, we have for all $r > |a_N|$

$$\sum_{n=1}^{N} \log \frac{r}{|a_n|} \leq \sum_{|a_n|<r} \log \frac{r}{|a_n|} \leq C - \log(|\alpha| r^m).$$

Letting r tend to 1, we find

$$0 < |\alpha| e^{-C} \leq \prod_{n=1}^{N} |a_n|,$$

so that $\sum (1 - |a_n|) < \infty$. To prove the converse, we have only to note that Jensen's theorem implies

$$\left(\prod_{|a_n|<r} |a_n| \right) \exp\left\{ \frac{1}{2\pi} \int_0^{2\pi} \log|f(re^{i\theta})|\, d\theta \right\} < |\alpha|.$$

COROLLARY. The zeros $\{z_n\}$ of a function of class N must satisfy

$$\sum (1 - |z_n|) < \infty.$$

One might well expect a stronger growth condition such as $f \in H^p$ to imply a correspondingly stronger statement about the rate at which the zeros approach the boundary. Even if $f \in H^\infty$, however, nothing more can be said about its zeros. In fact, given an arbitrary sequence $\{a_n\}$ of numbers in the unit disk such that $\sum (1 - |a_n|) < \infty$, one can produce a bounded analytic function whose zeros are precisely $\{a_n\}$. It is natural to construct such a function as an infinite product. The factors will be essentially the functions $(z - a_n)/(1 - \overline{a_n}z)$, which map the unit disk onto itself and vanish at a_n.

THEOREM 2.4. Let $a_1, a_2, \ldots$ be a sequence of complex numbers such that $0 < |a_1| \le |a_2| \le \cdots < 1$ and $\sum_{n=1}^\infty (1 - |a_n|) < \infty$. Then the infinite product

$$B(z) = \prod_{n=1}^\infty \frac{|a_n|}{a_n} \frac{a_n - z}{1 - \overline{a_n} z}$$

converges uniformly in each disk $|z| \le R < 1$. Each a_n is a zero of $B(z)$, with multiplicity equal to the number of times it occurs in the sequence; and $B(z)$ has no other zeros in $|z| < 1$. Finally, $|B(z)| < 1$ in $|z| < 1$, and $|B(e^{i\theta})| = 1$ a.e.

PROOF. In the disk $|z| \le R$,

$$\left| 1 - \frac{|a_n|}{a_n} \frac{a_n - z}{1 - \overline{a_n} z} \right| = \left| \frac{(a_n + |a_n|z)(1 - |a_n|)}{a_n(1 - \overline{a_n} z)} \right| \le \frac{2(1 - |a_n|)}{1 - R}.$$

Since $\sum (1 - |a_n|) < \infty$, it follows that the infinite product converges uniformly in each disk $|z| \le R < 1$, and therefore represents a function $B(z)$ analytic in $|z| < 1$. Furthermore, each a_n is a zero of $B(z)$ with the correct multiplicity, and $B(z) \neq 0$ otherwise. That $|B(z)| < 1$ in $|z| < 1$ is obvious since this is true for the partial products. The radial limit $B(e^{i\theta})$ therefore exists almost everywhere, and $|B(e^{i\theta})| \le 1$.

It remains to show that $|B(e^{i\theta})| = 1$ a.e. But for any $f \in H^\infty$, the Lebesgue bounded convergence theorem and the monotonicity of $M_1(r, f)$ give

$$\int_0^{2\pi} |f(re^{i\theta})| \, d\theta \le \int_0^{2\pi} |f(e^{it})| \, dt.$$

Now apply this to the function $f = B/B_n$, where

$$B_n(z) = \prod_{k=1}^n \frac{|a_k|}{a_k} \frac{a_k - z}{1 - \overline{a_k} z}.$$

Since $|B_n(e^{i\theta})| \equiv 1$, we have

$$\int_0^{2\pi} \left| \frac{B(re^{i\theta})}{B_n(re^{i\theta})} \right| d\theta \le \int_0^{2\pi} |B(e^{it})| \, dt.$$

But $B_n(z) \to B(z)$ uniformly on $|z| = r$, so

$$2\pi \leq \int_0^{2\pi} |B(e^{it})| \, dt.$$

Because $|B(e^{i\theta})| \leq 1$ a.e., this shows $|B(e^{i\theta})| = 1$ a.e.

A function of the form

$$B(z) = z^m \prod_n \frac{|a_n|}{a_n} \frac{a_n - z}{1 - \overline{a}_n z} \tag{1}$$

is called a *Blaschke product*. Here m is a nonnegative integer and $\sum (1 - |a_n|) < \infty$. The set $\{a_n\}$ may be finite, or even empty. If $\{a_n\}$ is empty, it is understood that $B(z) = z^m$.

2.3. MEAN CONVERGENCE TO BOUNDARY VALUES

We have seen that every H^p function $f(re^{i\theta})$ converges almost everywhere to an L^p boundary function $f(e^{i\theta})$. For the further development of the theory, it is important to know that $f(re^{i\theta})$ always tends to $f(e^{i\theta})$ in the sense of the L^p mean. The following factorization theorem will enter into the proof and will become a standard tool in later chapters.

THEOREM 2.5 (F. Riesz). Every function $f(z) \not\equiv 0$ of class H^p $(p > 0)$ can be factored in the form $f(z) = B(z)g(z)$, where $B(z)$ is a Blaschke product and $g(z)$ is an H^p function which does not vanish in $|z| < 1$. Similarly, each $f \in N$ has a factorization $f = Bg$, where g is a nonvanishing function of class N.

PROOF. We may suppose that $f(z)$ has infinitely many zeros, since otherwise the theorem is trivial. Let

$$B_n(z) = z^m \prod_{k=1}^n \frac{|a_k|}{a_k} \frac{a_k - z}{1 - \overline{a}_k z}$$

denote the partial Blaschke product, and let $g_n(z) = f(z)/B_n(z)$. For fixed n and $\varepsilon > 0$, $|B_n(z)| > 1 - \varepsilon$ for $|z|$ sufficiently close to 1. Therefore,

$$\int_0^{2\pi} |g_n(re^{i\theta})|^p \, d\theta \leq (1 - \varepsilon)^{-p} \int_0^{2\pi} |f(re^{i\theta})|^p \, d\theta \leq (1 - \varepsilon)^{-p} M$$

for r sufficiently large; hence, by monotonicity, for all r. Letting $\varepsilon \to 0$, we find

$$\int_0^{2\pi} |g_n(re^{i\theta})|^p \, d\theta \leq M$$

for all $r < 1$ and all n. By Theorem 2.4, however, $g_n(z)$ tends to $g(z) = f(z)/B(z)$ uniformly on each circle $|z| = R < 1$. Thus g is in H^p, and it has no zeros. (The proof for $f \in N$ is similar.)

We are now ready for the mean convergence theorem.

THEOREM 2.6. If $f \in H^p$ $(0 < p < \infty)$, then

$$\lim_{r \to 1} \int_0^{2\pi} |f(re^{i\theta})|^p \, d\theta = \int_0^{2\pi} |f(e^{i\theta})|^p \, d\theta \tag{2}$$

and

$$\lim_{r \to 1} \int_0^{2\pi} |f(re^{i\theta}) - f(e^{i\theta})|^p \, d\theta = 0. \tag{3}$$

PROOF. First let us prove (3) for $p = 2$. If $f(z) = \sum a_n z^n$ is in H^2, then $\sum |a_n|^2 < \infty$. But by Fatou's lemma,

$$\int_0^{2\pi} |f(re^{i\theta}) - f(e^{i\theta})|^2 \, d\theta \le \liminf_{\rho \to 1} \int_0^{2\pi} |f(re^{i\theta}) - f(\rho e^{i\theta})|^2 \, d\theta$$

$$= 2\pi \sum_{n=1}^\infty |a_n|^2 (1 - r^n)^2,$$

which tends to zero as $r \to 1$. This proves (3), and hence (2), in the case $p = 2$.

If $f \in H^p$ $(0 < p < \infty)$, we use the factorization $f = Bg$ given in Theorem 2.5. Since $[g(z)]^{p/2} \in H^2$, it follows from what we have just proved that

$$\int_0^{2\pi} |f(re^{i\theta})|^p \, d\theta \le \int_0^{2\pi} |g(re^{i\theta})|^p \, d\theta \to \int_0^{2\pi} |g(e^{i\theta})|^p \, d\theta = \int_0^{2\pi} |f(e^{i\theta})|^p \, d\theta.$$

This together with Fatou's lemma proves (2).

The following lemma can now be applied to deduce (3) from (2).

LEMMA 1. Let Ω be a measurable subset of the real line, and let $\varphi_n \in L^p(\Omega)$, $0 < p < \infty$; $n = 1, 2, \ldots$. As $n \to \infty$, suppose $\varphi_n(x) \to \varphi(x)$ a.e. on Ω and

$$\int_\Omega |\varphi_n(x)|^p \, dx \longrightarrow \int_\Omega |\varphi(x)|^p \, dx < \infty.$$

Then

$$\int_\Omega |\varphi_n(x) - \varphi(x)|^p \, dx \longrightarrow 0.$$

PROOF. For a measurable set $E \subset \Omega$, let $J_n(E) = \int_E |\varphi_n|^p$ and $J(E) = \int_E |\varphi|^p$. Let $\tilde{E} = \Omega - E$. Then

$$J(E) \leq \liminf_{n \to \infty} J_n(E) \leq \limsup_{n \to \infty} J_n(E)$$

$$\leq \lim_{n \to \infty} J_n(\Omega) - \liminf_{n \to \infty} J_n(\tilde{E}) \leq J(\Omega) - J(\tilde{E}) = J(E).$$

This shows that $J_n(E) \to J(E)$ for each $E \subset \Omega$.

Given $\varepsilon > 0$, choose a set $F \subset \Omega$ of finite measure such that $J(\tilde{F}) < \varepsilon$. Choose $\delta > 0$ so that $J(Q) < \varepsilon$ for every set $Q \subset F$ of measure $m(Q) < \delta$. By Egorov's theorem, there exists a set $Q \subset F$ with $m(Q) < \delta$, such that $\varphi_n(x) \to \varphi(x)$ uniformly on $E = F - Q$. Thus

$$\int_\Omega |\varphi_n - \varphi|^p = \int_F |\varphi_n - \varphi|^p + \int_Q |\varphi_n - \varphi|^p + \int_E |\varphi_n - \varphi|^p$$

$$\leq 2^p \{J_n(\tilde{F}) + J(\tilde{F}) + J_n(Q) + J(Q)\} + \int_E |\varphi_n - \varphi|^p$$

$$< (2^p \cdot 6 + 1)\varepsilon$$

for n sufficiently large, since $J_n(\tilde{F}) \to J(\tilde{F})$ and $J_n(Q) \to J(Q)$. This proves the lemma, and Theorem 2.6 follows.

COROLLARY. If $f \in H^p$ for some $p > 0$, then

$$\lim_{r \to 1} \int_0^{2\pi} \left| \log^+ |f(re^{i\theta})| - \log^+ |f(e^{i\theta})| \right| d\theta = 0.$$

The corollary is an immediate consequence of Theorem 2.6 and the following lemma.

LEMMA 2. For $a \geq 0$, $b \geq 0$, and for $0 < p \leq 1$,

$$|\log^+ a - \log^+ b| \leq (1/p) |a - b|^p.$$

PROOF. It is enough to assume $1 \leq b < a$. The result then follows from the inequality

$$\log x \leq (1/p) (x - 1)^p, \qquad x \geq 1,$$

by setting $x = a/b$. The latter inequality can be proved by noting that the difference of the two expressions has a positive derivative and vanishes at $x = 1$.

It is tempting to think that the corollary holds generally for all $f \in N$. This is *false*; the function $f(z) = \exp\{(1 + z)/(1 - z)\}$ is a counterexample.

There is also a "short" proof of the mean convergence theorem (Theorem 2.6) which does not use the Riesz factorization theorem but appeals instead to the Hardy–Littlewood maximal theorem (Theorem 1.9). If $f \in H^p$, then $f(re^{i\theta}) \to f(e^{i\theta})$ almost everywhere, and $f(e^{i\theta}) \in L^p$. But by the maximal theorem, $|f(re^{i\theta})| \leq F(\theta)$, where $F \in L^p$. Hence Theorem 2.6 follows at once from the Lebesgue dominated convergence theorem!

It should be observed at this point that H^p is a normed linear space if $1 \leq p \leq \infty$. The norm is defined as the L^p norm of the boundary function. Thus if $1 \leq p < \infty$,

$$\|f\|_p = \left\{ \frac{1}{2\pi} \int_0^{2\pi} |f(e^{i\theta})|^p \, d\theta \right\}^{1/p} = \lim_{r \to 1} M_p(r, f);$$

while

$$\|f\|_\infty = \sup_{|z|<1} |f(z)| = \operatorname{ess\ sup}_{0 \leq \theta < 2\pi} |f(e^{i\theta})|.$$

2.4. CANONICAL FACTORIZATION

The Riesz factorization (Theorem 2.5) can be refined to produce a canonical factorization which is of supreme importance both for the theory of H^p spaces and for its applications. This refinement rests upon the following inequality.

THEOREM 2.7. If $f \in H^p$, $p > 0$, then

$$\log|f(re^{i\theta})| \leq \frac{1}{2\pi} \int_0^{2\pi} P(r, \theta - t) \log|f(e^{it})| \, dt.$$

PROOF. After factoring out the Blaschke product, whose presence would only strengthen the inequality, we may assume $f(z) \neq 0$ in $|z| < 1$. Then $\log|f(z)|$ is harmonic in $|z| < 1$, and

$$\log|f(\rho e^{i\theta})| = \frac{1}{2\pi} \int_0^{2\pi} P(r, \theta - t) \log|f(\rho e^{it})| \, dt, \qquad r < \rho < 1.$$

By the corollary to Theorem 2.6,

$$\lim_{\rho \to 1} \int_0^{2\pi} P(r, \theta - t) \log^+|f(\rho e^{it})| \, dt = \int_0^{2\pi} P(r, \theta - t) \log^+|f(e^{it})| \, dt.$$

On the other hand, by Fatou's lemma,

$$\lim_{\rho \to 1} \int_0^{2\pi} P(r, \theta - t) \log^-|f(\rho e^{it})| \, dt \geq \int_0^{2\pi} P(r, \theta - t) \log^-|f(e^{it})| \, dt.$$

Subtraction gives the desired result.

The theorem is false for the class N, as the example $\exp\{(1 + z)/(1 - z)\}$ again shows. The reciprocal of this function reveals, incidentally, that strict inequality may occur even if $f(z)$ is bounded and has no zeros.

Returning to the problem of factorization, let $f(z) \not\equiv 0$ be of class H^p for some $p > 0$. According to Theorem 2.2, $f(e^{i\theta}) \in L^p$ and $\log|f(e^{i\theta})| \in L^1$. Consider the analytic function

$$F(z) = \exp\left\{\frac{1}{2\pi} \int_0^{2\pi} \frac{e^{it} + z}{e^{it} - z} \log|f(e^{it})|\, dt\right\}. \tag{4}$$

Let $f(z) = B(z)g(z)$ as in Theorem 2.5; thus $g(z) \neq 0$ and $|g(e^{i\theta})| = |f(e^{i\theta})|$ a.e. By Theorem 2.7, $|g(z)| \leq |F(z)|$ in $|z| < 1$. Also, $|g(e^{i\theta})| = |F(e^{i\theta})|$ a.e., by Corollary 2 to Theorem 1.2. Hence if $e^{i\gamma} = g(0)/|g(0)|$, the function $S(z) = e^{-i\gamma}g(z)/F(z)$ is analytic in $|z| < 1$ and has the properties

$$0 < |S(z)| \leq 1; \qquad |S(e^{i\theta})| = 1 \text{ a.e.}; \qquad S(0) > 0.$$

This shows that $-\log|S(z)|$ is a positive harmonic function which vanishes almost everywhere on the boundary. Thus by the Herglotz representation and Theorem 1.2, $-\log|S(z)|$ can be represented as a Poisson–Stieltjes integral with respect to a bounded nondecreasing function $\mu(t)$, and $\mu'(t) = 0$ a.e. Since $S(0) > 0$, analytic completion gives

$$S(z) = \exp\left\{-\int_0^{2\pi} \frac{e^{it} + z}{e^{it} - z}\, d\mu(t)\right\}. \tag{5}$$

Putting everything together, we have the factorization $f(z) = e^{i\gamma}B(z)S(z)F(z)$.

We now introduce some terminology. An *outer function* for the class H^p is a function of the form

$$F(z) = e^{i\gamma} \exp\left\{\frac{1}{2\pi} \int_0^{2\pi} \frac{e^{it} + z}{e^{it} - z} \log \psi(t)\, dt\right\}, \tag{6}$$

where γ is a real number, $\psi(t) \geq 0$, $\log \psi(t) \in L^1$, and $\psi(t) \in L^p$. Thus (4) is an outer function.

An *inner function* is any function $f(z)$ analytic in $|z| < 1$, having the properties $|f(z)| \leq 1$ and $|f(e^{i\theta})| = 1$ a.e. We have shown that every inner function has a factorization $e^{i\gamma}B(z)S(z)$, where $B(z)$ is a Blaschke product and $S(z)$ is a function of the the form (5), $\mu(t)$ being a bounded nondecreasing singular function ($\mu'(t) = 0$ a.e.). Such a function $S(z)$ is called a *singular inner function*.

THEOREM 2.8 (Canonical factorization theorem). Every function $f(z) \not\equiv 0$ of class H^p ($p > 0$) has a unique factorization of the form $f(z) = B(z)S(z)F(z)$, where $B(z)$ is a Blaschke product, $S(z)$ is a singular inner function, and $F(z)$

is an outer function for the class H^p (with $\psi(t) = |f(e^{it})|$). Conversely, every such product $B(z)S(z)F(z)$ belongs to H^p.

PROOF. We have already shown that every $f \in H^p$ can be factored as claimed, and the uniqueness is obvious. To prove the converse, it suffices to show that an outer function (6) must belong to H^p. Applying the arithmetic–geometric mean inequality (see Exercise 2), we find

$$|F(z)|^p \leq \frac{1}{2\pi} \int_0^{2\pi} P(r, \theta - t)[\psi(t)]^p \, dt.$$

Thus

$$\int_0^{2\pi} |F(re^{i\theta})|^p \, d\theta \leq \int_0^{2\pi} [\psi(t)]^p \, dt.$$

There is a similar factorization for $f \in N$. A function $F(z)$ of the form (6), where $\psi(t) \geq 0$ and $\log \psi(t) \in L^1$, will be called an *outer function* for the class N. (Note that the condition $\psi \in L^p$ has been dropped.)

THEOREM 2.9. Every function $f(z) \not\equiv 0$ of class N can be expressed in the form

$$f(z) = B(z)[S_1(z)/S_2(z)]F(z), \tag{7}$$

where $B(z)$ is a Blaschke product, $S_1(z)$ and $S_2(z)$ are singular inner functions, and $F(z)$ is an outer function for the class N (with $\psi(t) = |f(e^{it})|$). Conversely, every function of the form (7) belongs to N.

PROOF. Let $f(z) = e^{i\gamma}B(z)g(z)$, where $g \in N$, $g(z) \neq 0$ in $|z| < 1$, and $g(0) > 0$. Since $\log|g(z)| \in h^1$, it has a representation as a Poisson–Stieltjes integral

$$\log|g(z)| = \frac{1}{2\pi} \int_0^{2\pi} P(r, \theta - t) \, d\nu(t)$$

with respect to a function $\nu(t)$ of bounded variation. Analytic completion and separation of $\nu(t)$ into its absolutely continuous and singular components gives the desired representation. The converse follows directly from the fact that every Poisson–Stieltjes integral is of class h^1.

2.5. THE CLASS N^+

The two preceding theorems point out the sharp structural difference between functions in the classes H^p and N. In factoring functions of class N, it is necessary not only to enlarge the class of admissible outer functions, but also to replace the singular factor by a *quotient* of two singular inner functions.

This allows, for instance, our "pathological" example $\exp\{(1 + z)/(1 - z)\}$, which we now recognize as the reciprocal of a singular inner function. It is useful to distinguish the class N^+ of all functions $f \in N$ for which $S_2(z) \equiv 1$. That is, $f \in N^+$ if it has the form $f = BSF$, where B is a Blaschke product, S is a singular inner function, and F is an outer function for the class N. In a sense, N^+ is the natural limit of H^p as $p \to 0$. The proper inclusions $H^p \subset N^+ \subset N$ are obvious.

THEOREM 2.10. A function $f \in N$ belongs to the class N^+ if and only if

$$\lim_{r \to 1} \int_0^{2\pi} \log^+ |f(re^{i\theta})| \, d\theta = \int_0^{2\pi} \log^+ |f(e^{i\theta})| \, d\theta. \tag{8}$$

PROOF. Suppose first that $f \in N^+$, so that $f = BSF$. Then, in view of (4),

$$\log^+ |f(re^{i\theta})| \leq \log^+ |F(re^{i\theta})| \leq \frac{1}{2\pi} \int_0^{2\pi} P(r, \theta - t) \log^+ |f(e^{it})| \, dt.$$

Hence

$$\lim_{r \to 1} \int_0^{2\pi} \log^+ |f(re^{i\theta})| \, d\theta \leq \int_0^{2\pi} \log^+ |f(e^{it})| \, dt.$$

Fatou's lemma gives the reverse inequality.

It is more difficult to prove the sufficiency of the condition (8). We first observe that for an arbitrary Blaschke product $B(z)$,

$$\lim_{r \to 1} \int_0^{2\pi} \log |B(re^{i\theta})| \, d\theta = 0. \tag{9}$$

This is obvious if $B(z)$ has only a finite number of factors, so we may assume

$$B(z) = \prod_{n=1}^{\infty} \frac{|a_n|}{a_n} \frac{a_n - z}{1 - \bar{a}_n z}, \qquad 0 < |a_n| \leq |a_{n+1}| < 1.$$

By Jensen's theorem,

$$\frac{1}{2\pi} \int_0^{2\pi} \log |B(re^{i\theta})| \, d\theta = \sum_{|a_n| < r} \log \frac{r}{|a_n|} + \log |B(0)|.$$

Holding N fixed, we then have for all $r > |a_N|$

$$\frac{1}{2\pi} \int_0^{2\pi} \log |B(re^{i\theta})| \, d\theta \geq \sum_{n=1}^{N} \log \frac{r}{|a_n|} + \sum_{n=1}^{\infty} \log |a_n|$$

$$= N \log r + \sum_{n=N+1}^{\infty} \log |a_n|.$$

Consequently,

$$2\pi \sum_{n=N+1}^{\infty} \log|a_n| \le \lim_{r \to 1} \int_0^{2\pi} \log|B(re^{i\theta})| \, d\theta \le 0,$$

and (9) follows by letting $N \to \infty$.

Continuing the proof of the theorem, let the given function $f \in N$ be expressed in the form $f(z) = B(z)g(z)$, where $g \in N$ and $g(z) \ne 0$ in $|z| < 1$. Since

$$\log|B(z)| + \log^+|g(z)| \le \log^+|f(z)| \le \log^+|g(z)|,$$

combination of (8) and (9) gives

$$\lim_{r \to 1} \int_0^{2\pi} \log^+|g(re^{i\theta})| \, d\theta = \int_0^{2\pi} \log^+|g(e^{i\theta})| \, d\theta. \tag{10}$$

Since $\log|g(z)| \in h^1$, it has a representation

$$\log|g(z)| = \frac{1}{2\pi} \int_0^{2\pi} P(r, \theta - t) \, dv(t) \tag{11}$$

with respect to a function $v(t)$ of bounded variation. Recalling the proof of Theorem 1.1, we see that v can be chosen to have the form

$$v(\theta) = \lim_{n \to \infty} \int_0^{\theta} \log|g(r_n e^{it})| \, dt, \qquad 0 \le \theta \le 2\pi,$$

where $\{r_n\}$ is an appropriate sequence increasing to 1. On the other hand, by Fatou's lemma,

$$\int_0^{\theta} \log^+|g(e^{it})| \, dt \le \liminf_{n \to \infty} \int_0^{\theta} \log^+|g(r_n e^{it})| \, dt = v_+(\theta), \tag{12}$$

say. If there were strict inequality for some θ, then a similar application of Fatou's lemma in $[\theta, 2\pi]$ and addition of the two results would give a contradiction to (10). Equality therefore holds in (12) for all θ, which shows that $v_+(\theta)$ is absolutely continuous. On the other hand,

$$v_-(\theta) = v_+(\theta) - v(\theta) = \liminf_{n \to \infty} \int_0^{\theta} \log^-|g(r_n e^{it})| \, dt$$

is nondecreasing. In view of (11), this shows that $g = SG$, where S is a singular inner function and G is an outer function. Hence $f \in N^+$, which was to be shown.

The following useful result is an easy consequence of the factorization theorems.

THEOREM 2.11. If $f \in N^+$ and $f(e^{i\theta}) \in L^p$ for some $p > 0$, then $f \in H^p$.

The *a priori* assumption that $f \in N^+$ cannot be relaxed. The reciprocal of any (nontrivial) singular inner function is bounded on $|z| = 1$, but is not of class N^+.

2.6. HARMONIC MAJORANTS

We noted in Section 1.3 that if $f(z)$ is analytic in a domain D, then $|f(z)|^p$ is subharmonic in D. This means that in each disk contained in D, $|f(z)|^p$ is dominated by a harmonic function, the Poisson integral of its boundary function. However, there may not be a *single* harmonic function which dominates $|f(z)|^p$ throughout D. In general, a function $g(z)$ is said to have a *harmonic majorant* in D if there is a function $U(z)$ harmonic in D such that $g(z) \le U(z)$ for all z in D. If g is continuous and has a harmonic majorant, it is obviously subharmonic; but the converse is false.

THEOREM 2.12. If $f(z)$ is analytic in $|z| < 1$, then $f \in H^p$ $(0 < p < \infty)$ if and only if $|f(z)|^p$ has a harmonic majorant in $|z| < 1$.

PROOF. If $U(z)$ is a harmonic majorant of $|f(z)|^p$, then by the mean value theorem

$$M_p(r, f) \le [U(0)]^{1/p}.$$

Conversely, if $f \in H^p$, it follows from Theorem 2.7 and the arithmetic–geometric mean inequality (Exercise 2) that

$$|f(z)|^p \le \exp\left\{ \frac{1}{2\pi} \int_0^{2\pi} P(r, \theta - t) \log[|f(e^{it})|^p]\, dt \right\}$$

$$\le \frac{1}{2\pi} \int_0^{2\pi} P(r, \theta - t)|f(e^{it})|^p\, dt.$$

In other words, $|f(z)|^p$ is dominated by the Poisson integral $U(z)$ of its boundary function.

It is easy to see, in fact, that this is the *least* harmonic majorant of $|f(z)|^p$. That is, if $V(z)$ is any other harmonic majorant, then $U(z) \le V(z)$ for all z, $|z| < 1$. Indeed, for any $\rho < 1$.

$$\frac{1}{2\pi} \int_0^{2\pi} P(r, \theta - t)|f(\rho e^{it})|^p\, dt \le \frac{1}{2\pi} \int_0^{2\pi} P(r, \theta - t)V(\rho e^{it})\, dt$$

$$= V(\rho z).$$

As $\rho \to 1$, this shows $U(z) \le V(z)$.

COROLLARY. If $f \in H^p$ $(0 < p < \infty)$ and if φ is analytic and satisfies $|\varphi(z)| < 1$ in $|z| < 1$, then $g(z) = f(\varphi(z)) \in H^p$ and

$$M_p(r, g) \leq \left\{ \frac{1 + |\varphi(0)|}{1 - |\varphi(0)|} \right\}^{1/p} M_p(1, f).$$

PROOF. Let $U(z)$ be the Poisson integral of $|f(e^{it})|^p$. Then $|f(z)|^p \leq U(z)$, so that $U(\varphi(z))$ is a harmonic majorant of $|g(z)|^p$. Hence the mean value theorem gives

$$\{M_p(r, g)\}^p \leq U(\varphi(0)) \leq \frac{1 + |\varphi(0)|}{1 - |\varphi(0)|} \{M_p(1, f)\}^p,$$

since $P(r, \theta) \leq (1 + r)/(1 - r)$.

EXERCISES

1. Prove in detail that $f \in N$ if and only if $\int \left| \log |f(re^{i\theta})| \right| d\theta$ is bounded.

2. (Arithmetic–geometric mean inequality.) Let $f(x) \geq 0$ be integrable with respect to a nonnegative measure $d\mu$ of unit total mass. Let $\mathfrak{A}(f) = \int f(x) \, d\mu(x)$ be the "arithmetic mean" of f, and let $\mathfrak{G}(f) = \exp\{\mathfrak{A}(\log f)\}$ be the "geometric mean." Prove that $\mathfrak{G}(f) \leq \mathfrak{A}(f)$, with equality if and only if $f(x)$ is constant a.e. (*Hint*: Note that $\log t \leq t - 1$, $t \geq 0$, and set $t = f(x)/\mathfrak{A}(f)$. This proof is due to F. Riesz [7].)

3. Show that the classical inequality

$$(a_1 a_2 \cdots a_n)^{1/n} \leq (1/n) \, (a_1 + a_2 + \cdots + a_n), \qquad a_k > 0,$$

is a special case of the general arithmetic–geometric mean inequality.

4. (Jensen's inequality) Let $\varphi(u)$ be a convex function in an interval $a < u < b$, and suppose $a < f(x) < b$. As in Exercise 2, let $\mathfrak{A}(f)$ denote the arithmetic mean of f with respect to a unit measure $d\mu$. Prove Jensen's inequality $\varphi(\mathfrak{A}(f)) \leq \mathfrak{A}(\varphi(f))$, and deduce the general arithmetic–geometric mean inequality. (*Hint*: Let $\alpha = \mathfrak{A}(f)$ and let λ be the slope of a supporting line through $(\alpha, \varphi(\alpha))$. Show that

$$\varphi(u) - \varphi(\alpha) \geq \lambda(u - \alpha), \qquad a < u < b.$$

This proof is due to Zygmund [4].) Interpret Jensen's inequality for the case in which the unit measure $d\mu$ is concentrated at a finite number of points.

5. Show that if $\psi(u)$ is a nondecreasing convex function and $g(z)$ is subharmonic, then $\psi(g(z))$ is subharmonic.

6. Show that a function $f(z)$ analytic in $|z| < 1$ is a Blaschke product (aside from a constant factor of modulus one) if and only if

$$\lim_{r \to 1} \int_0^{2\pi} \left|\log|f(re^{i\theta})|\right| d\theta = 0.$$

7. Show that if $f(z)$ is an inner function and $\varphi(z)$ is a conformal mapping of the unit disk onto itself, then $f(\varphi(z))$ and $\varphi(f(z))$ are inner functions.

8. Let $f(z)$ be an inner function, and suppose $|\alpha| < 1$. Prove that for all α outside a set of planar measure zero, the function

$$\frac{f(z) - \alpha}{1 - \bar{\alpha}f(z)}$$

is a Blaschke product! (Frostman [1].)

9. Prove the corollary to Theorem 2.6 under the weaker hypothesis that $f \in N^+$.

10. Show that for functions f analytic in the unit disk, $f \in N$ if and only if $\log^+|f(z)|$ has a harmonic majorant.

NOTES

Fatou [1] showed that each $f \in H^\infty$ has a nontangential limit $f(e^{i\theta})$ almost everywhere, and that $f(e^{i\theta})$ cannot vanish on an arc $\theta_0 \le \theta \le \theta_0 + \delta$ unless $f(z) \equiv 0$. F. and M. Riesz [1] improved this to H^1 and showed that $f(z) \equiv 0$ if its boundary function vanishes on a set of positive measure. Later, Szegö [1] showed that if $f \in H^2$ and $f(z) \not\equiv 0$, then $\log|f(e^{i\theta})| \in L^1$. After F. Riesz [3] proved his factorization theorem (Theorem 2.5), he was able to show that each $f \in H^p$ $(p > 0)$ has a radial limit almost everywhere, and that $\log|f(e^{i\theta})| \in L^1$ unless $f(z) \equiv 0$. He also proved the mean convergence of $f(re^{i\theta})$ to $f(e^{i\theta})$ (Theorem 2.6). Lemma 1, which simplifies the proof of this theorem, is also due to F. Riesz [6], at least for $p \ge 1$. The proof given in the text is in Littlewood's book [5]. A somewhat different approach to the structure of H^p functions is in the paper of M. and G. Weiss [1]. For further information on the boundary behavior of H^p functions, see Tanaka [1].

Theorem 2.1 is due to F. and R. Nevanlinna [1]; it unifies the presentation of the H^p theory. A paper of Ostrowski [1] is also relevant here. Blaschke [1] introduced "Blaschke products" and proved Theorem 2.4. Regarding matters of priority, however, see Landau [3]. The canonical factorization theorems (Theorems 2.8 and 2.9) are due to Smirnov [2], who also noted a weaker form of Theorem 2.11: if $f \in H^p$ and $f(e^{i\theta}) \in L^q$ for some $q > p$, then

$f \in H^q$. Beurling [1] coined the terms "inner function" and "outer function." Smirnov [3] introduced the class N^+ (called "D" in his paper and in subsequent Soviet literature) and cited Theorem 2.10, which he attributed to "Madame P. Kotchine". More about the class N^+ may be found in Privalov's book [4]. The condition (9) actually characterizes Blaschke products among all analytic functions with $|f(z)| \le 1$ (see Exercise 6). This theorem is due to M. Riesz, but was first published in Frostman [1]. (See also Zygmund [4], Vol. I, p. 281.) Theorem 2.12 was mentioned in passing by Smirnov ([3], p. 341). There is a large literature on the boundary behavior of Blaschke products; see, for example, Frostman [2], Cargo [1, 2], and Somadasa [1].

Paley and Zygmund [2] have shown that if the hypothesis $f \in N$ is "slightly relaxed" (which they interpret in two different ways), then $f(z)$ need not have a radial limit on any set of positive measure.

We now turn to some applications of the H^p theory to such diverse topics as measure theory (the F. and M. Riesz theorem), Cauchy and Poisson integrals, and conformal mapping. Some of the results obtained will be useful in the later development of the H^p theory. Further applications will appear in subsequent chapters.

3.1. POISSON INTEGRALS AND H^1

In Chapter 1, the harmonic functions $u \in h^1$ were characterized as Poisson–Stieltjes integrals. We shall now show that if the harmonic conjugate of u also belongs to h^1, then the representing function $\mu(t)$ is absolutely continuous. This result will have a number of interesting consequences.

NOTE. The harmonic conjugate of u is any function v such that $f(z) = u(z) + iv(z)$ is analytic in the disk. We shall speak of " the " harmonic conjugate even though it is determined only up to an additive constant. Later it will be convenient to normalize v by the requirement $v(0) = 0$.

THEOREM 3.1. A function $f(z)$ analytic in $|z| < 1$ is representable in the form

$$f(z) = \frac{1}{2\pi} \int_0^{2\pi} P(r, \theta - t)\varphi(t)\, dt \tag{1}$$

as the Poisson integral of a function $\varphi \in L^1$ if and only if $f \in H^1$. In this case, $\varphi(t) = f(e^{it})$ a.e.

PROOF. If an analytic function $f(z)$ has the form (1), then

$$\int_0^{2\pi} |f(re^{i\theta})|\, d\theta \le \int_0^{2\pi} |\varphi(t)|\, dt,$$

so that $f \in H^1$. Conversely, suppose $f \in H^1$, and write

$$\Phi(z) = \frac{1}{2\pi} \int_0^{2\pi} P(r, \theta - t) f(e^{it})\, dt.$$

For any fixed $\rho, 0 < \rho < 1$,

$$f(\rho z) = \frac{1}{2\pi} \int_0^{2\pi} P(r, \theta - t) f(\rho e^{it})\, dt.$$

But, by Theorem 2.6, $\int |f(\rho e^{it}) - f(e^{it})|\, dt \to 0$ as $\rho \to 1$, so $f(\rho z) \to \Phi(z)$. Hence $\Phi(z) = f(z)$, and the theorem is proved.

COROLLARY 1. Let $u \in h^1$, so that it is the Poisson–Stieltjes integral of a function $\mu(t)$ of bounded variation. If the conjugate harmonic function $v \in h^1$, then $\mu(t)$ is absolutely continuous.

COROLLARY 2. A function $f(z)$ analytic in $|z| < 1$ is the Poisson integral of a function $\varphi \in L^p$ $(1 \le p \le \infty)$ if and only if $f \in H^p$.

COROLLARY 3. Let $f(z) = B(z)[S_1(z)/S_2(z)]F(z) \in N$. If $\log[f(z)/B(z)] \in H^1$, then $S_1(z)/S_2(z) \equiv 1$.

Corollary 2 follows from Theorem 2.11.

We shall prove in Chapter 4 that the harmonic conjugate of any function $u \in h^p$ $(1 < p < \infty)$ is also of class h^p. This cannot be true for $p = 1$, as Corollary 1 shows, even if u is a positive harmonic function. In fact, the Poisson kernel $P(r, \theta)$ itself is a counterexample. The following weaker theorem, however, remains true.

THEOREM 3.2. Every analytic function $f(z)$ with positive real part in $|z| < 1$ is of class H^p for all $p < 1$.

PROOF. Without loss of generality, suppose $f(0) = 1$. The range of f is contained in the right half-plane, so f is subordinate to

$$\frac{1 + z}{1 - z} = P(r, \theta) + iQ(r, \theta),$$

where $P(r, \theta)$ is the Poisson kernel and

$$Q(r, \theta) = \frac{2r \sin \theta}{1 - 2r \cos \theta + r^2}$$

is the conjugate Poisson kernel. It follows from Littlewood's subordination theorem (Theorem 1.7) that

$$\int_0^{2\pi} |f(re^{i\theta})|^p \, d\theta \leq \int_0^{2\pi} \left| \frac{1 + re^{i\theta}}{1 - re^{i\theta}} \right|^p \, d\theta \leq \int_0^{2\pi} |Q(1, \theta)|^p \, d\theta < \infty$$

for any $p < 1$, since $(1 + z)/(1 - z)$ is in H^p and $P(1, \theta) = 0$ for $\theta \neq 0$.

COROLLARY. If $u \in h^1$, then its harmonic conjugate belongs to h^p for all $p < 1$.

This follows from the representation of h^1 functions as the difference of two positive harmonic functions (Theorem 1.1). In Chapter 4 the corollary will be reproved in slightly stronger form (Kolmogorov's theorem).

3.2. DESCRIPTION OF THE BOUNDARY FUNCTIONS

We have seen that each function $f \in H^p$ ($0 < p \leq \infty$) has a nontangential limit $f(e^{i\theta})$ at almost every boundary point. It is of interest to characterize these boundary functions in a simple way. Let $\mathscr{H}^p$ denote the set of boundary functions $f(e^{i\theta})$ of functions $f \in H^p$. As usual, two functions are identified if they coincide almost everywhere, so the elements of $\mathscr{H}^p$ are actually equivalence classes. We know that $\mathscr{H}^p \subset L^p$, and in fact that $\mathscr{H}^p$ is a vector subspace of L^p: it is closed under addition and scalar multiplication. We shall see presently that $\mathscr{H}^p$ is also topologically closed. For $\varphi \in L^p$, let

$$\|\varphi\| = \|\varphi\|_p = \left\{ \frac{1}{2\pi} \int_0^{2\pi} |\varphi(t)|^p \, dt \right\}^{1/p}.$$

It is convenient to retain the norm notation even for $p < 1$, when $\| \ \|_p$ is not a genuine norm. The notation $\|f\|_p$ for $f \in H^p$ will refer to the boundary function. Thus

$$\|f\| = \|f\|_p = M_p(1, f) = \lim_{r \to 1} M_p(r, f).$$

LEMMA. If $f \in H^p$ $(0 < p < \infty)$, then

$$|f(z)| \leq 2^{1/p} \|f\|_p (1 - r)^{-1/p}, \qquad r = |z|.$$

PROOF. If $p = 1$, the estimate follows easily from the Poisson integral representation of H^1 functions (Theorem 3.1). For $p \neq 1$, we may use the Riesz factorization (Theorem 2.5) of a function $f \in H^p$ in the form $f = Bg$, where B is a Blaschke product and g is a nonvanishing H^p function. Since $|f(z)| \leq |g(z)|$ and $\|f\|_p = \|g\|_p$, it suffices to prove the lemma for non-vanishing H^p functions. But if $g \in H^p$ and $g(z) \neq 0$, some branch of $[g(z)]^p$ is single-valued and belongs to H^1. Thus the desired result follows from the H^1 case.

We are now ready to characterize $\mathscr{H}^p$. It is clear that $\mathscr{H}^p$ contains all functions of the form

$$\sum_{k=0}^{n} a_k e^{ik\theta},$$

where the a_k are complex constants. These functions will be called *polynomials in $e^{i\theta}$*. They are dense in $\mathscr{H}^p$, as we shall now show.

THEOREM 3.3. For $0 < p < \infty$, $\mathscr{H}^p$ is the L^p closure of the set of polynomials in $e^{i\theta}$.

PROOF. For $0 < \rho < 1$, let $f_\rho(z) = f(\rho z)$. Given $f \in H^p$ and $\varepsilon > 0$, choose $\rho < 1$ such that

$$\|f_\rho - f\|_p < \varepsilon/2.$$

(This is possible, by Theorem 2.6.) Now let $S_n(z)$ denote the nth partial sum of the Taylor series of f at the origin. Since $S_n(z) \to f(z)$ uniformly on the circle $|z| = \rho$,

$$\|S_{n_\rho} - f_\rho\|_p < \varepsilon/2$$

for n sufficiently large. Thus, using Minkowski's inequality in the case $p \geq 1$, we find

$$\|S_{n_\rho} - f\|_p < \varepsilon.$$

This shows that the boundary function $f(e^{i\theta})$ belongs to the L^p closure of the polynomials in $e^{i\theta}$. For $p < 1$, the inequality

$$(a + b)^p \leq 2^p(a^p + b^p), \qquad a > 0, \quad b > 0,$$

gives the same result.

To complete the proof, we have to show that $\mathscr{H}^p$ is closed. But suppose that a sequence $\{f_n(e^{i\theta})\}$ of $\mathscr{H}^p$ functions converges in L^p mean to $\varphi(\theta) \in L^p$. Then by the lemma, $\{f_n(z)\}$ is uniformly bounded in each disk $|z| \leq R < 1$. Thus $\{f_n\}$ is a normal family, so a diagonalization argument produces a subsequence $\{f_{n_k}(z)\}$ which converges uniformly in each disk $|z| \leq R < 1$ to an analytic function $f(z)$. It is clear that $f \in H^p$. We wish to show that $\varphi(\theta) = f(e^{i\theta})$ a.e. But given $\varepsilon > 0$, choose N such that $\|f_n - f_m\|_p < \varepsilon$ for $n, m \geq N$. Then for $m \geq N$ and $r < 1$,

$$M_p(r, f - f_m) = \lim_{k \to \infty} M_p(r, f_{n_k} - f_m) \leq \lim_{k \to \infty} \sup \|f_{n_k} - f_m\|_p < \varepsilon.$$

Letting $r \to 1$, we find $\|f - f_m\|_p < \varepsilon$ for all $m \geq N$; thus $\|f - f_n\|_p \to 0$, and $\varphi(\theta) = f(e^{i\theta})$ a.e.

The same argument, in simpler form, shows that $\mathscr{H}^\infty$ is closed.

COROLLARY 1. *If $1 \leq p \leq \infty$, H^p is a Banach space.*

If $p < 1$, then $\| \ \|_p$ is not a true norm and in fact the space H^p is not normable. However, the expression

$$d(f, g) = \|f - g\|_p{}^p$$

defines a metric on H^p if $p < 1$. This can be verified using the inequality

$$(a + b)^p \leq a^p + b^p, \qquad a > 0, \quad b > 0,$$

which is valid for $0 < p < 1$. (See Section 4.2.) Furthermore, the theorem shows that H^p is complete under the topology induced by this metric.

COROLLARY 2. *If $0 < p < 1$, H^p is a complete metric space.*

Theorem 3.3 is false for $p = \infty$, since $\mathscr{H}^\infty$ contains functions which do not coincide almost everywhere with continuous functions. One example is the singular inner function $\exp\{(z + 1)/(z - 1)\}$, whose boundary function is $\exp\{-i \cot(\theta/2)\}$, $\theta \neq 0$.

There is another approach, however, which leads to a description of $\mathscr{H}^p$ for $1 \leq p \leq \infty$. The *Fourier coefficients* of a function $\varphi \in L^1$ are the numbers

$$c_n = \int_0^{2\pi} e^{-int} \varphi(t)\, dt, \qquad n = 0, \pm 1, \pm 2, \dots .$$

It is important to note that the Taylor coefficients of a function $f \in H^p$ $(1 \leq p \leq \infty)$ coincide with the Fourier coefficients of its boundary function. The following theorem expresses this more precisely.

THEOREM 3.4. Let $f(z) = \sum_{n=0}^{\infty} a_n z^n$ belong to H^1, and let $\{c_n\}$ be the Fourier coefficients of its boundary function $f(e^{it})$. Then $c_n = a_n$ for $n \geq 0$, and $c_n = 0$ for $n < 0$. Furthermore, $\mathscr{H}^p$ $(1 \leq p \leq \infty)$ is exactly the class of L^p functions whose Fourier coefficients vanish for all $n < 0$.

PROOF. The Taylor coefficients of f can be expressed in the form

$$a_n = \frac{1}{2\pi} \int_0^{2\pi} r^{-n} e^{-int} f(re^{it})\, dt, \qquad 0 < r < 1.$$

Thus

$$|r^n a_n - c_n| \leq \|f_r - f\|_1 \to 0$$

as $r \to 1$, by Theorem 2.6. [Here, as above, $f_r(z) = f(rz)$.] This shows that $c_n = a_n$ for $n \geq 0$. Similarly, $c_n = 0$ for $n < 0$.

Suppose now that $\varphi \in L^p$ $(1 \leq p \leq \infty)$ and that its Fourier coefficients c_n vanish for all $n < 0$. Let

$$f(z) = \frac{1}{2\pi} \int_0^{2\pi} P(r, \theta - t)\varphi(t)\, dt, \qquad z = re^{i\theta},$$

be the Poisson integral of φ. Since the Poisson kernel has the expansion

$$P(r, t) = 1 + \sum_{n=1}^{\infty} r^n (e^{int} + e^{-int}),$$

it follows that

$$f(z) = \sum_{n=0}^{\infty} c_n z^n, \qquad |z| < 1.$$

This shows that $f(z)$ is analytic in $|z| < 1$; hence $f \in H^p$, by Corollary 2 to Theorem 3.1. But $\varphi(t) = f(e^{it})$ a.e., so $\varphi \in \mathscr{H}^p$, and the proof is complete.

According to the Riemann–Lebesgue lemma, the Fourier coefficients of an integrable function tend to zero as $n \to \pm\infty$. Thus we have the following corollary.

COROLLARY. If $f(z) = \sum a_n z^n \in H^1$, then $a_n \to 0$ as $n \to \infty$.

From this point on we will usually suppress the notation $\mathscr{H}^p$ for the space of boundary functions. It is convenient, and customary in the literature, to identify $\mathscr{H}^p$ with H^p, and thus to regard H^p as a subspace of L^p.

3.3. CAUCHY AND CAUCHY–STIELTJES INTEGRALS

For each complex-valued function $\varphi(\zeta)$ integrable over the unit circle $|\zeta| = 1$, the *Cauchy integral*

$$F(z) = \frac{1}{2\pi i} \int_{|\zeta|=1} \frac{\varphi(\zeta)\, d\zeta}{\zeta - z} = \frac{1}{2\pi} \int_0^{2\pi} \frac{e^{it}\varphi(e^{it})}{e^{it} - z}\, dt$$

represents a pair of functions: one analytic in $|z| < 1$, the other in $|z| > 1$. More generally, the same is true for a *Cauchy–Stieltjes integral*

$$F(z) = \frac{1}{2\pi} \int_0^{2\pi} \frac{e^{it}\, d\mu(t)}{e^{it} - z}$$

with respect to a complex-valued function $\mu(t)$ of bounded variation.

THEOREM 3.5. If a function $F(z)$ can be represented in $|z| < 1$ as a Cauchy–Stieltjes integral, then it belongs to H^p for all $p < 1$.

PROOF. It suffices to assume $\mu(t)$ is real and nondecreasing. Then $\mathrm{Re}\{F(z)\} > 0$, since

$$\mathrm{Re}\left(\frac{e^{it}}{e^{it} - z}\right) = \frac{1 - r\cos(t - \theta)}{1 - 2r\cos(t - \theta) + r^2} > 0;$$

and the result follows from Theorem 3.2.

The converse is false: a function can be in H^p for all $p < 1$, yet not be representable as a Cauchy–Stieltjes integral. (See Exercise 2.) We shall also see that a Cauchy–Stieltjes integral, or even a Cauchy integral, can easily fail to be of class H^1.

A Cauchy–Stieltjes integral is related in a simple way to the Poisson–Stieltjes integral formed with respect to the same function $\mu(t)$:

$$F(z) - F\left(\frac{1}{\bar{z}}\right) = \frac{1}{2\pi} \int_0^{2\pi} P(r, \theta - t)\, d\mu(t), \qquad |z| < 1. \tag{2}$$

In view of Theorem 3.5 and Theorem 2.2, the radial limit

$$\lim_{r \to 1-} F(re^{i\theta})$$

exists almost everywhere (taken from inside the unit circle). But as $r \to 1$, the right-hand side of (2) tends to $\mu'(\theta)$ a.e., by Theorem 1.2. Thus the exterior radial limit also exists a.e., and

$$\lim_{r \to 1-} F(re^{i\theta}) - \lim_{r \to 1+} F(re^{i\theta}) = \mu'(\theta) \quad \text{a.e.}$$

THEOREM 3.6. Every function $f \in H^1$ can be expressed as the Cauchy integral of its boundary function. In fact,

$$f^{(k)}(z) = \frac{k!}{2\pi i} \int_{|\zeta|=1} \frac{f(\zeta)\, d\zeta}{(\zeta - z)^{k+1}}, \qquad |z| < 1; \quad k = 0, 1, 2, \ldots .$$

Each of these integrals vanishes identically in $|z| > 1$.

PROOF. It suffices to establish the result for $k = 0$; the rest will then follow by the familiar procedure of "differentiation under the integral sign." (See, for example, Ahlfors [2], p. 121.) Let $F(z)$ be the Cauchy integral of $f(e^{it})$, and let $\{P_n(z)\}$ be a sequence of polynomials such that $\|f - P_n\|_1 \to 0$. (See Theorem 3.3.) Writing

$$F(z) = \frac{1}{2\pi i} \int_{|\zeta|=1} \frac{f(\zeta) - P_n(\zeta)}{\zeta - z}\, d\zeta, \qquad |z| > 1,$$

we find that $F(z) \equiv 0$ in $|z| > 1$. Hence it follows from the relation (2) with $d\mu(t) = f(e^{it})\, dt$, and from Theorem 3.1, that

$$F(z) = \frac{1}{2\pi} \int_0^{2\pi} P(r, \theta - t) f(e^{it})\, dt = f(z), \qquad |z| < 1. \tag{3}$$

This completes the proof.

The relation (2) can also be used to prove the following theorem.

THEOREM 3.7. For a complex-valued function $\mu(t)$ of bounded variation, the following three statements are equivalent:

(i) $$\int_0^{2\pi} e^{int}\, d\mu(t) = 0, \qquad n = 1, 2, \ldots .$$

(ii) The Cauchy–Stieltjes integral

$$F(z) = \frac{1}{2\pi} \int_0^{2\pi} \frac{e^{it}\, d\mu(t)}{e^{it} - z}$$

vanishes identically in $|z| > 1$.

(iii) The Poisson–Stieltjes integral

$$f(z) = \frac{1}{2\pi} \int_0^{2\pi} P(r, \theta - t)\, d\mu(t)$$

is analytic in $|z| < 1$.

PROOF. (i) $\Leftrightarrow$ (ii). This follows by expanding the kernel $e^{it}(e^{it} - z)^{-1}$ in powers of $1/z$ and integrating term by term.

(ii) ⇒ (iii). Since the Cauchy–Stieltjes integral $F(z)$ is always analytic in $|z| < 1$, this follows at once from (2).

(iii) ⇒ (ii). The Cauchy–Stieltjes integral $F(z)$ is analytic in $|z| > 1$. If the Poisson–Stieltjes integral $f(z)$ is analytic in $|z| < 1$, then (2) shows that $F(\bar{z})$ is also analytic in $|z| > 1$. Thus by the Cauchy–Riemann equations, $F(z)$ is identically constant in $|z| > 1$. Since $F(z) \to 0$ as $z \to \infty$, the constant must be zero. Hence $F(z) \equiv 0$ in $|z| > 1$.

A complex-valued function $\mu(t)$ of bounded variation over $[0, 2\pi]$ will be called *normalized* if $\mu(t) = \mu(t+)$ for $0 \le t < 2\pi$; that is, if μ is continuous from the right. We can now use the preceding results to establish an important measure-theoretic theorem.

THEOREM 3.8 (F. and M. Riesz). Let $\mu(t)$ be a normalized complex-valued function of bounded variation on $[0, 2\pi]$, with the property

$$\int_0^{2\pi} e^{int} \, d\mu(t) = 0, \qquad n = 1, 2, \ldots .$$

Then $\mu(t)$ is absolutely continuous.

PROOF. By Theorem 3.7, the hypothesis implies the corresponding Poisson–Stieltjes integral $f(z)$ is analytic in $|z| < 1$; hence $f \in H^1$. But according to Theorem 3.1, every H^1 function can be represented in the form

$$f(z) = \frac{1}{2\pi} \int_0^{2\pi} P(r, \theta - t) f(e^{it}) \, dt.$$

Since the Poisson representation is unique (Section 1.2), $d\mu(t) = f(e^{it}) \, dt$.

The essential step in the proof was the use of Theorem 3.1, the representation of H^1 functions by Poisson integrals. In fact, this result is in a sense *equivalent* to the theorem of F. and M. Riesz. Below is another variant which is sometimes useful. The proof is left as an exercise.

THEOREM 3.9. If a function $f(z)$ analytic in $|z| < 1$ can be represented in any one of the following four ways:
   (i) as a Cauchy–Stieltjes integral with $F(z) \equiv 0$ in $|z| > 1$;
   (ii) as a Cauchy integral with $F(z) \equiv 0$ in $|z| > 1$;
   (iii) as a Poisson–Stieltjes integral;
   (iv) as a Poisson integral,
then it can be represented in each of the other three ways. The class of functions so representable is H^1.

3.4. ANALYTIC FUNCTIONS CONTINUOUS IN $|z| \leq 1$

In general, a continuous function of bounded variation need not be absolutely continuous. For example, the familiar Lebesgue function over the Cantor set is monotonic, continuous, and purely singular. For the boundary values of functions analytic in the disk, however, continuity is equivalent to absolute continuity. In fact, a somewhat stronger statement can be made, as follows.

THEOREM 3.10. If $f \in H^1$ and its boundary function $f(e^{i\theta})$ is equal almost everywhere to a function of bounded variation, then $f(z)$ is continuous in $|z| \leq 1$ and $f(e^{i\theta})$ is absolutely continuous.

PROOF. Suppose $f(e^{i\theta}) = \mu(\theta)$ a.e., where μ is a normalized function of bounded variation and $\mu(0) = \mu(2\pi)$. By Theorem 3.4,

$$\int_0^{2\pi} e^{in\theta} \mu(\theta)\, d\theta = \int_0^{2\pi} e^{in\theta} f(e^{i\theta})\, d\theta = 0, \qquad n = 1, 2, \dots.$$

Integrating by parts, we find

$$\int_0^{2\pi} e^{in\theta}\, d\mu(\theta) = 0, \qquad n = 1, 2, \dots.$$

By the F. and M. Riesz theorem, then, $\mu(\theta)$ is absolutely continuous. On the other hand, $f(z)$ can be represented as a Poisson integral of $\mu(\theta)$, by Theorem 3.1 and the fact that $f(e^{i\theta}) = \mu(\theta)$ a.e. Thus the radial limit $f(e^{i\theta})$ exists *everywhere*, coincides with $\mu(\theta)$, and is absolutely continuous.

We shall now show that the functions just considered—analytic functions with absolutely continuous boundary values—are characterized by the simple condition $f' \in H^1$. The expression $f'(e^{i\theta})$ then can have two possible meanings. It may indicate the radial limit of $f'(z)$; or it may indicate the derivative with respect to θ of the boundary function, apart from a factor $ie^{i\theta}$. It is a remarkable fact that the two are essentially the same.

THEOREM 3.11. A function $f(z)$ analytic in $|z| < 1$ is continuous in $|z| \leq 1$ and absolutely continuous on $|z| = 1$ if and only if $f' \in H^1$. If $f' \in H^1$, then

$$\frac{d}{d\theta} f(e^{i\theta}) = ie^{i\theta} \lim_{r \to 1} f'(re^{i\theta}) \quad \text{a.e.} \tag{4}$$

PROOF. If $f(z)$ is analytic in $|z| < 1$ and continuous in $|z| \leq 1$, it is the Poisson integral of its boundary function:

$$f(z) = \frac{1}{2\pi} \int_0^{2\pi} P(r, \theta - t) f(e^{it})\, dt.$$

Differentiate with respect to θ:

$$izf'(z) = \frac{1}{2\pi} \int_0^{2\pi} \frac{\partial P}{\partial \theta}(r, \theta - t) f(e^{it}) \, dt;$$

and integrate by parts [using the absolute continuity of $f(e^{it})$] to obtain

$$izf'(z) = \frac{1}{2\pi} \int_0^{2\pi} P(r, \theta - t) i e^{it} f'(e^{it}) \, dt. \tag{5}$$

Thus $izf'(z) \in H^1$, which implies $f' \in H^1$.

Conversely, if $f' \in H^1$, then $izf'(z)$ can be represented in the form (5) as the Poisson integral of its boundary function. Here it is understood that $f'(e^{it})$ denotes $\lim_{r \to 1} f'(re^{it})$. Since the function

$$g(\theta) = \int_0^\theta i e^{it} f'(e^{it}) \, dt$$

is absolutely continuous in $[0, 2\pi]$, and $g(0) = g(2\pi) = 0$, integration by parts in (5) gives

$$\frac{\partial}{\partial \theta} \{f(re^{i\theta})\} = \frac{\partial}{\partial \theta} \left\{ \frac{1}{2\pi} \int_0^{2\pi} P(r, \theta - t) g(t) \, dt \right\}.$$

Thus

$$f(re^{i\theta}) = \frac{1}{2\pi} \int_0^{2\pi} P(r, \theta - t) g(t) \, dt + C(r).$$

Since $C(r)$ is the difference of two (complex-valued) harmonic functions, it is itself harmonic:

$$C''(r) + (1/r)C'(r) = 0.$$

Thus $C(r) = a \log r + b$, where a and b are constants. To ensure continuity at the origin, a must be zero. Hence $f(z)$ is the Poisson integral of the continuous function $[g(t) + b]$, and is therefore continuous in $|z| \leq 1$ and has boundary values $f(e^{i\theta}) = g(\theta) + b$. The relation (4) now follows from the definition of $g(\theta)$.

3.5. APPLICATIONS TO CONFORMAL MAPPING

The theorems just given, together with a few general facts about H^p functions, can be applied to obtain some rather deep results in the theory of conformal mapping.

A *Jordan curve* (or a *simple closed curve*) C is the image of a continuous complex-valued function $w = w(t)$ $(0 \leq t \leq 2\pi)$ such that $w(0) = w(2\pi)$ and

$w(t_1) \neq w(t_2)$ for $0 \leq t_1 < t_2 < 2\pi$. The curve C is said to be *rectifiable* if $w(t)$ is of bounded variation. Its length L is then defined as the total variation of $w(t)$:

$$L = \sup \sum_{k=1}^{n} |w(t_k) - w(t_{k-1})|,$$

where the supremum is taken over all finite partitions

$$0 = t_0 < t_1 < \cdots < t_n = 2\pi$$

of $[0, 2\pi]$. It is easily seen that L depends only on the curve C, and is invariant under a change of parameter. If $w(t)$ is absolutely continuous, the length of the arc of C corresponding to an arbitrary interval $a \leq t \leq b$ is given by

$$\int_a^b |w'(t)| \, dt.$$

(A proof may be found in Natanson [1], Vol. II, p. 227.) It follows that any Lebesgue measurable set $E \subset [0, 2\pi]$ has an image on C of measure

$$\int_E |w'(t)| \, dt.$$

Indeed, the formula is obviously valid for open sets E, hence for G_δ sets (countable intersections of open sets). Therefore, since an arbitrary measurable set E is contained in a G_δ set with the same measure, the formula holds generally.

A famous theorem of Carathéodory asserts that every conformal mapping $w = f(z)$ of $|z| < 1$ onto the interior of a Jordan curve C has a one–one continuous extension to $|z| \leq 1$. In particular, $w = f(e^{i\theta})$ is a parametrization of C. Applying Theorems 3.10 and 3.11, we therefore obtain the following important result.

THEOREM 3.12. Let $f(z)$ map $|z| < 1$ conformally onto the interior of a Jordan curve C. Then C is rectifiable if and only if $f' \in H^1$.

This theorem is highly plausible, perhaps even "obvious," when viewed geometrically. It says that the lengths

$$L_r = r \int_0^{2\pi} |f'(re^{i\theta})| \, d\theta$$

of the images of the circles $|z| = r$ are bounded if and only if the boundary has finite length.

In the presence of a rectifiable boundary, this result puts the general H^p theory at our disposal. The derivative of the mapping function has an angular limit $f'(e^{i\theta})$ almost everywhere on the boundary, and $\log|f'(e^{i\theta})|$ is integrable. The function $f'(e^{i\theta})$ is related as in (4) to the derivative of the absolutely continuous boundary function $f(e^{i\theta})$. A measurable set E on the unit circle is carried onto a subset of C with measure

$$\int_E |f'(e^{i\theta})| \, d\theta.$$

Consequently, a subset of $|z| = 1$ has measure zero if and only if its image on C has measure zero. In other words, the boundary sets of measure zero are preserved under the conformal mapping. This remains true (in view of the Riemann mapping theorem) for a conformal mapping between any two Jordan domains with rectifiable boundaries.

More can be said. If $f(z)$ is a conformal mapping of $|z| < 1$ onto the interior of a rectifiable Jordan curve C, then its continuous extension to $|z| \leq 1$ is conformal at almost every boundary point. To be precise, let γ be a continuous curve in $|z| < 1$ which terminates at a point $z_0 = e^{i\theta_0}$ in a well-defined direction not tangent to the unit circle; i.e., the limit of $\arg\{z - z_0\}$ is to exist as $z \to z_0$ along γ, and is not to equal $\theta_0 \pm \pi/2$. The image of γ is then a curve Γ inside C, terminating at $f(z_0)$. Since $(d/d\theta)\{f(e^{i\theta})\}$ exists a.e., C has a tangent direction at almost every point. We assert that for almost every θ_0, the angle between Γ and the tangent to C at $f(z_0)$ exists and is equal to the angle between γ and the tangent to the unit circle at z_0. In other words, the mapping preserves angles at almost every boundary point.

We have to show that

$$\arg\left\{\left[\frac{d}{d\theta}f(e^{i\theta})\right]_{\theta=\theta_0}\right\} - \lim_{\substack{z \to z_0 \\ z \in \gamma}} \arg\{f(z) - f(z_0)\}$$

$$= \theta_0 + \pi/2 - \lim_{\substack{z \to z_0 \\ z \in \gamma}} \arg\{z - z_0\}.$$

In view of relation (4), it is enough to prove

$$\lim_{\substack{z \to z_0 \\ z \in \gamma}} \frac{f(z) - f(z_0)}{z - z_0} = f'(z_0) \tag{6}$$

wherever the angular limit $f'(z_0)$ of $f'(z)$ exists; hence almost everywhere. But

$$\frac{f(z) - f(z_0)}{z - z_0} = \frac{1}{z - z_0} \int_{z_0}^{z} f'(\zeta) \, d\zeta, \tag{7}$$

the integration being performed along the segment joining z_0 and z. If $f'(z_0)$ exists, the right-hand side of (7) approaches $f'(z_0)$ as $z \to z_0$ along γ, which was assumed to be a nontangential path. This proves (6).

3.6. INEQUALITIES OF FEJÉR–RIESZ, HILBERT, AND HARDY

We shall now discuss some interesting inequalities which will have applications in later chapters.

THEOREM 3.13 (Fejér–Riesz inequality). If $f \in H^p$ $(0 < p < \infty)$, then the integral of $|f(x)|^p$ along the segment $-1 \leq x \leq 1$ converges, and

$$\int_{-1}^{1} |f(x)|^p \, dx \leq \tfrac{1}{2} \int_{0}^{2\pi} |f(e^{i\theta})|^p \, d\theta. \tag{8}$$

The constant $\tfrac{1}{2}$ is best possible.

PROOF. Consider first the case $p = 2$, and suppose for the moment that $f(z)$ is real on the real axis. By Cauchy's theorem with a semi-circular path,

$$\int_{-r}^{r} [f(x)]^2 \, dx + i \int_{0}^{\pi} [f(re^{i\theta})]^2 e^{i\theta} \, d\theta = 0$$

for each $r < 1$. It follows that

$$\int_{-r}^{r} [f(x)]^2 \, dx \leq \int_{0}^{\pi} |f(re^{i\theta})|^2 \, d\theta.$$

Adding a similar inequality for the lower semicircle, we find

$$2 \int_{-r}^{r} [f(x)]^2 \, dx \leq \int_{0}^{2\pi} |f(re^{i\theta})|^2 \, d\theta \leq \int_{0}^{2\pi} |f(e^{i\theta})|^2 \, d\theta.$$

The desired inequality is now obtained by letting r tend to 1.

More generally, we may express an arbitrary H^2 function $f(z)$ in the form

$$f(z) = \sum_{n=0}^{\infty} (\alpha_n + i\beta_n)z^n = \sum_{n=0}^{\infty} \alpha_n z^n + i \sum_{n=0}^{\infty} \beta_n z^n = g(z) + ih(z),$$

where g and h are in H^2 and are real on the real axis. By what we have just proved,

$$\int_{-1}^{1} |f(x)|^2 \, dx = \int_{-1}^{1} |g(x)|^2 \, dx + \int_{-1}^{1} |h(x)|^2 \, dx$$

$$\leq \frac{1}{2} \int_{0}^{2\pi} |g(e^{i\theta})|^2 \, d\theta + \frac{1}{2} \int_{0}^{2\pi} |h(e^{i\theta})|^2 \, d\theta$$

$$= \frac{1}{2} \int_{0}^{2\pi} |f(e^{i\theta})|^2 \, d\theta - \frac{i}{2} \int_{0}^{2\pi} [h(e^{i\theta})g(e^{-i\theta}) - g(e^{i\theta})h(e^{-i\theta})] \, d\theta.$$

But the last integral must vanish, since the integrand is an odd function of θ. Hence inequality (8) is established in the case $p = 2$.

To deduce the result for general p, we use a familiar device. If $f(z)$ belongs to H^p and vanishes nowhere in $|z| < 1$, then $[f(z)]^{p/2} \in H^2$, and (8) follows from the special case $p = 2$ already proved. If $f(z)$ has zeros, we factor out the Blaschke product $B(z)$ to obtain a nonvanishing function $g(z) = f(z)/B(z)$ which is again of class H^p. Thus

$$\int_{-1}^{1} |f(x)|^p \, dx \leq \int_{-1}^{1} |g(x)|^p \, dx \leq \tfrac{1}{2} \int_{0}^{2\pi} |g(e^{i\theta})|^p \, d\theta = \tfrac{1}{2} \int_{0}^{2\pi} |f(e^{i\theta})|^p \, d\theta.$$

In view of Theorem 3.12, there is an interesting geometric application.

COROLLARY. If the unit disk $|z| < 1$ is mapped conformally onto the interior of a rectifiable Jordan curve C, the image of any diameter has length at most half the length of C.

We may now finish the proof of the theorem by showing that $\tfrac{1}{2}$ is the best possible constant. Let $w = \varphi(z)$ map $|z| < 1$ conformally onto the interior of the rectangle with vertices $\pm 1 \pm i\varepsilon$, the diameter $-1 \leq z \leq 1$ corresponding to the real segment $-1 \leq w \leq 1$. The ratio of the length of this segment to the perimeter of the rectangle is $2[4(1 + \varepsilon)]^{-1}$, which tends to $\tfrac{1}{2}$ as $\varepsilon \to 0$. The function $f(z) = [\varphi'(z)]^{1/p} \in H^p$ therefore shows that the constant cannot be improved.

Two further inequalities, named after Hilbert and Hardy, lie in the same circle of ideas. We shall prove them in generalized form, with a view to later applications. For a complex vector $x = (x_0, x_1, \ldots, x_N)$, let

$$\|x\|^2 = \sum_{n=0}^{N} |x_n|^2.$$

THEOREM 3.14. Let $\psi \in L^\infty$ and

$$\lambda_n = \frac{1}{2\pi} \int_{0}^{2\pi} e^{-int} \psi(t) \, dt, \qquad n = 0, 1, 2, \ldots. \tag{9}$$

Let

$$A_N(x, y) = \sum_{n, m=0}^{N} \lambda_{n+m} x_n y_m.$$

Then

$$|A_N(x, y)| \leq \|\psi\|_\infty \|x\| \|y\|.$$

PROOF. Let $P(t) = \sum_{n=0}^{N} x_n e^{-int}$. Then

$$|A_N(x, x)| = \left| \frac{1}{2\pi} \int_{0}^{2\pi} [P(t)]^2 \psi(t) \, dt \right|$$

$$\leq \|\psi\|_\infty \frac{1}{2\pi} \int_{0}^{2\pi} |P(t)|^2 \, dt = \|\psi\|_\infty \|x\|^2.$$

To deduce the more general result, observe that

$$A_N(x, y) = \tfrac{1}{4}A_N(x + y, x + y) - \tfrac{1}{4}A_N(x - y, x - y).$$

Thus

$$|A_N(x, y)| \le \tfrac{1}{4}\|\psi\|_\infty \{\|x + y\|^2 + \|x - y\|^2\}$$
$$= \tfrac{1}{2}\|\psi\|_\infty \{\|x\|^2 + \|y\|^2\}.$$

This shows that $|A_N(x, y)| \le \|\psi\|_\infty$ if $\|x\| = \|y\| = 1$, which is enough to prove the theorem.

COROLLARY (Hilbert's inequality).

$$\left| \sum_{n, m=0}^{N} \frac{x_n y_m}{n + m + 1} \right| \le \pi \|x\| \, \|y\|.$$

PROOF. Choose $\psi(t) = ie^{-it}(\pi - t)$, so that $\lambda_n = (n + 1)^{-1}$ and $\|\psi\|_\infty = \pi$.

THEOREM 3.15. Let $\lambda_n \ge 0$ be given by (9) for some $\psi \in L^\infty$. Then if $f(z) = \sum a_n z^n \in H^1$,

$$\sum_{n=0}^{\infty} \lambda_n |a_n| \le \|\psi\|_\infty \|f\|_1.$$

PROOF. Every $f \in H^1$ has a factorization $f = gh$, where g and h are H^2 functions with $\|g\|_2^2 = \|h\|_2^2 = \|f\|_1$. Indeed, $f = B\varphi$, where B is a Blaschke product and φ is a nonvanishing H^1 function (Theorem 2.5). Let $g = B\varphi^{1/2}$ and $h = \varphi^{1/2}$. Now let $g(z) = \sum b_n z^n$ and $h(z) = \sum c_n z^n$; then

$$a_n = \sum_{k=0}^{n} b_k c_{n-k}.$$

Hence

$$\sum_{n=0}^{N} \lambda_n |a_n| \le \sum_{n=0}^{N} \lambda_n \sum_{k=0}^{n} |b_k| \, |c_{n-k}|$$

$$\le \sum_{k, m=0}^{N} \lambda_{k+m} |b_k| \, |c_m| \le \|\psi\|_\infty \|g\|_2 \|h\|_2 = \|\psi\|_\infty \|f\|_1,$$

by Theorem 3.14.

COROLLARY (Hardy's inequality). If $f(z) = \sum a_n z^n \in H^1$, then

$$\sum_{n=0}^{\infty} \frac{|a_n|}{n + 1} \le \pi \|f\|_1.$$

Hardy's inequality is proved with the same choice of ψ that gave Hilbert's inequality. One interesting consequence should be mentioned. Suppose $f(z) = \sum a_n z^n$ is analytic in $|z| < 1$. If $\sum |a_n| < \infty$, then f has a continuous extension to $|z| \leq 1$, but the converse is false (see Exercise 7). Hardy's inequality shows, however, that if $f' \in H^1$ (or equivalently, in light of Theorem 3.11, if f is continuous in $|z| \leq 1$ and *absolutely* continuous on $|z| = 1$), then $\sum |a_n| < \infty$. In particular, $\sum |a_n| < \infty$ if f is a conformal mapping of the unit disk onto a Jordan domain with rectifiable boundary.

3.7. SCHLICHT FUNCTIONS

A function analytic in a domain is said to be *schlicht* (or *univalent*) if it does not take any value twice; that is, if $f(z_1) \neq f(z_2)$ whenever $z_1 \neq z_2$. Our main aim in this section is to prove that every schlicht function in the unit disk is of class H^p for all $p < \frac{1}{2}$. This will show, for instance, that a conformal mapping of the unit disk onto an arbitrary simply connected domain, however pathological the boundary, automatically has a finite radial limit in almost every direction. The Koebe function

$$k(z) = \frac{z}{(1-z)^2} = \sum_{n=1}^{\infty} nz^n,$$

which maps $|z| < 1$ onto the full plane slit along the negative real axis from $-\frac{1}{4}$ to ∞, shows that a schlicht function need not belong to $H^{1/2}$.

We say that $f \in \mathscr{S}$ if f is analytic and schlicht in $|z| < 1$ and is normalized so that $f(0) = 0$ and $f'(0) = 1$. The first of the following lemmas is very well known, and the proof is omitted here. (See, for example, Hayman [1], p. 4.)

LEMMA 1. If $f \in \mathscr{S}$, then

$$\frac{r}{(1+r)^2} \leq |f(re^{i\theta})| \leq \frac{r}{(1-r)^2}.$$

LEMMA 2. Let C_1 and C_2 be twice continuously differentiable Jordan curves in the plane, surrounding the origin, with C_1 in the interior of C_2. Then

$$\int_{C_1} r^p \, d\theta \leq \int_{C_2} r^p \, d\theta, \qquad p > 0,$$

where (r, θ) are polar coordinates and both curves are traversed in the positive direction.

PROOF. Let D be the ring domain between C_1 and C_2, let C be its boundary, and let (x, y) denote rectangular coordinates. Applying Green's theorem and the Cauchy–Riemann equations

$$\frac{\partial(\log r)}{\partial x} = \frac{\partial \theta}{\partial y}; \qquad \frac{\partial(\log r)}{\partial y} = -\frac{\partial \theta}{\partial x},$$

we find

$$\int_C r^p \, d\theta = p \iint_D r^{p-1} \left[\frac{\partial r}{\partial x} \frac{\partial \theta}{\partial y} - \frac{\partial r}{\partial y} \frac{\partial \theta}{\partial x} \right] dx \, dy$$

$$= p \iint_D r^p \left[\left(\frac{\partial \theta}{\partial x}\right)^2 + \left(\frac{\partial \theta}{\partial y}\right)^2 \right] dx \, dy \geq 0.$$

This proves the lemma.

We are now ready to state the main theorem.

THEOREM 3.16. If $f(z)$ is analytic and schlicht in $|z| < 1$, then $f \in H^p$ for all $p < 1/2$.

PROOF. Without loss of generality, we may assume $f \in \mathscr{S}$, so that $f(z) \neq 0$ unless $z = 0$. Now set

$$f(re^{i\theta}) = Re^{i\Phi} \qquad (r > 0)$$

and apply the Cauchy–Riemann equation

$$r \frac{\partial(\log R)}{\partial r} = \frac{\partial \Phi}{\partial \theta}$$

to obtain

$$\frac{d}{dr} \int_0^{2\pi} |f(re^{i\theta})|^p \, d\theta = \int_0^{2\pi} \frac{\partial}{\partial r} (R^p) \, d\theta$$

$$= \frac{p}{r} \int_0^{2\pi} R^p \frac{\partial \Phi}{\partial \theta} \, d\theta = \frac{p}{r} \int_{\Gamma_r} R^p \, d\Phi,$$

where Γ_r is the image under f of the circle $|z| = r$. Thus if $M(r) = M_\infty(r, f)$ is the maximum of $|f(z)|$ on $|z| = r$, then for each $\varepsilon > 0$ the circle $|w| = M(r) + \varepsilon$ surrounds Γ_r, and it follows from Lemma 2 that

$$\frac{d}{dr} \{M_p{}^p(r, f)\} \leq \frac{p}{r} [M(r) + \varepsilon]^p.$$

Now let $\varepsilon \to 0$, integrate from 0 to r, and apply Lemma 1:

$$M_p{}^p(r, f) \leq p \int_0^r \frac{[M(r)]^p}{r} \, dr$$

$$\leq p \int_0^1 r^{p-1}(1-r)^{-2p} \, dr < \infty$$

if $0 < p < \frac{1}{2}$. This concludes the proof.

As a function of class $H^p(p < \frac{1}{2})$, each schlicht function f has a canonical factorization

$$f(z) = B(z)S(z)F(z)$$

(see Theorem 2.8). The Blaschke product $B(z)$ obviously has at most one factor. Less obvious is the fact that f can have no singular part.

THEOREM 3.17. If $f(z)$ is analytic and schlicht in $|z| < 1$, then its singular factor $S(z) \equiv 1$.

PROOF. If f does not vanish in the open disk, then $1/f$ is analytic and schlicht, so $1/f \in H^p$ for all $p < \frac{1}{2}$. Hence by the uniqueness of the canonical factorization of a function of class N (Theorem 2.9), $S(z) \equiv 1$.
If $f(\zeta) = 0$ for some ζ, $|\zeta| < 1$, then the function

$$g(z) = (1 - |\zeta|^2)^{-1}[f'(\zeta)]^{-1} f\left(\frac{z + \zeta}{1 + \bar{\zeta}z}\right)$$

belongs to $\mathscr{S}$. By Lemma 1, then,

$$|g(z)/z| \geq (1 + r)^{-2} > \tfrac{1}{4}.$$

From this it follows that $B/f \in H^\infty$, where

$$B(z) = \frac{z - \zeta}{1 - \bar{\zeta}z}.$$

In particular, f cannot have a (nontrivial) singular factor.

EXERCISES

1. Prove: If $f(z)$ is analytic and $\mathrm{Re}\{f(z)\} > 0$ in $|z| < 1$, then f is an outer function. (*Hint*: Show that $f_\varepsilon(z) = \varepsilon + f(z)$ is outer and let $\varepsilon \to 0$.)

2. Show that

$$f(z) = \frac{1}{1 - z} \log\left(\frac{1}{1 - z}\right)$$

belongs to H^p for all $p < 1$, but has unbounded Taylor coefficients. In particular, show that $f(z)$ is not a Cauchy–Stieltjes integral, so that the converse to Theorem 3.4 is false.

3. Prove Theorem 3.9.

4. Prove the integral analogue of Hilbert's inequality:

$$\left| \int_0^\infty \int_0^\infty \frac{f(x)g(y)}{x+y} \, dx \, dy \right| \le \pi \|f\|_2 \|g\|_2$$

if f and g are in $L^2(0, \infty)$.

5. It is a natural conjecture that if $f \in N$ and $f(e^{i\theta}) \in L^1$, then

$$\int_0^{2\pi} e^{in\theta} f(e^{i\theta}) \, d\theta = 0, \qquad n = 1, 2, \ldots .$$

Show this is false.

6. Construct an analytic function $f(z) = \sum a_n z^n$ such that $\sum |a_n| < \infty$, but $f' \notin H^1$.

7. Give an example of a function $f(z) = \sum a_n z^n$ analytic in $|z| < 1$ and continuous in $|z| \le 1$, such that $\sum |a_n| = \infty$. (*Suggestion*: Let f map the unit disk conformally onto a Jordan domain constructed in such a way that the radius $0 \le z < 1$ corresponds to a curve of infinite length.)

NOTES

Theorems 3.1 and 3.8 are essentially due to F. and M. Riesz [1]. They also showed that the coefficients of an H^1 function tend to zero. Theorem 3.4 and the falsity of its converse (Exercise 2) may be found in Smirnov [2]. Theorem 3.9 is due to Fichtenholz [1]. Privalov's book [4] discusses these results. Helson [1] has given a "soft" proof of the F. and M. Riesz theorem, making no use of complex function theory. Rudin [6] obtained a generalization of this theorem in which $\int e^{int} \, d\mu = 0$ only outside a certain "thin" set of positive integers. Theorem 3.11 is due to Privalov [1, 2], as are most of the results in Section 3.5. Theorems 3.10 and 3.12 appear in a paper of Smirnov [4]. The theorem of Carathéodory mentioned before Theorem 3.12 may be found in Goluzin [3], Chap II, Sec. 3, or in Zygmund [4], Chap. VII, Sec. 10. Golubev [1] gave the first example of a conformal mapping of the unit disk onto a Jordan domain which carries a boundary set of measure zero onto a set of positive measure on the (nonrectifiable) boundary of the image domain. Theorem 3.13 is in a paper of Fejér and Riesz [1]. Hardy [1] had essentially proved it for $p = 2$. Many proofs of Hilbert's inequality have been given; see Hardy, Littlewood, and Pólya [1]. "Hardy's inequality" seems to have appeared first in a paper of Hardy and Littlewood [1]. Theorem 3.16 is essentially in a paper of Prawitz [1]; see also Goluzin [3], Chap. IV, Sec. 6. Another proof is in Littlewood [5]. Theorem 3.17 is due to Lohwater and Ryan [1].

If a harmonic function has a certain property, must the same be true of its conjugate? Questions of this kind have been widely investigated, both for their own interest and for their importance in applications. In this chapter and the next, we consider growth and smoothness properties of functions harmonic in the unit disk. Generally speaking, a harmonic function and its conjugate behave alike, but there are some rather surprising exceptions.

4.1. THEOREM OF M. RIESZ

Given a real-valued function $u(z)$ harmonic in $|z| < 1$, let $v(z)$ be its harmonic conjugate, normalized so that $v(0) = 0$. Thus $f(z) = u(z) + iv(z)$ is analytic in $|z| < 1$, and $f(0)$ is real. If $f(z) = \sum c_n z^n$ and $c_n = a_n - ib_n$, then

$$u(z) = a_0 + \sum_{n=1}^{\infty} r^n(a_n \cos n\theta + b_n \sin n\theta),$$

$$v(z) = \sum_{n=1}^{\infty} r^n(-b_n \cos n\theta + a_n \sin n\theta).$$

(1)

By the Parseval relation,

$$[M_2(r, u)]^2 = a_0{}^2 + \frac{1}{2} \sum_{n=1}^{\infty} r^{2n}(a_n{}^2 + b_n{}^2),$$

$$[M_2(r, v)]^2 = \frac{1}{2} \sum_{n=1}^{\infty} r^{2n}(a_n{}^2 + b_n{}^2).$$

Hence

$$M_2(r, v) \le M_2(r, u), \tag{2}$$

with equality if and only if $a_0 = 0$. It was the discovery of M. Riesz that something similar is true for every p in the range $1 < p < \infty$.

THEOREM 4.1 (M. Riesz). If $u \in h^p$ for some p, $1 < p < \infty$, then its harmonic conjugate v is also of class h^p. Furthermore, there is a constant A_p, depending only on p, such that

$$M_p(r, v) \le A_p M_p(r, u), \qquad 0 \le r < 1, \tag{3}$$

for all $u \in h^p$.

PROOF. First let us observe that only the case $1 < p \le 2$ needs to be considered. Indeed, if the theorem holds for some index p $(1 < p < \infty)$, then it is also true for the "conjugate index" q defined by $1/p + 1/q = 1$, and in fact with $A_q = A_p$. To show this, let

$$g(z) = g(re^{i\theta}) = a_0 + \sum_{n=1}^{N} r^n(a_n \cos n\theta + b_n \sin n\theta)$$

be an arbitrary harmonic polynomial, and let

$$h(z) = \sum_{n=1}^{N} r^n(-b_n \cos n\theta + a_n \sin n\theta)$$

be the conjugate function, normalized by $h(0) = 0$. Thus $g(z) + ih(z)$ is an ordinary algebraic polynomial; and for fixed ρ $(0 < \rho < 1)$, the function

$$F(z) = [u(\rho z) + iv(\rho z)][g(z) + ih(z)]$$

is analytic in $|z| \le 1$. The mean value theorem applied to $\text{Im}\{F(z)\}$ gives

$$\int_0^{2\pi} [u(\rho e^{i\theta})h(e^{i\theta}) + v(\rho e^{i\theta})g(e^{i\theta})] \, d\theta = 2\pi[u(0)h(0) + v(0)g(0)] = 0.$$

Assuming the theorem has been proved for the index p, it follows that

$$\left| \frac{1}{2\pi} \int_0^{2\pi} v(\rho e^{i\theta})g(e^{i\theta}) \, d\theta \right| = \left| \frac{1}{2\pi} \int_0^{2\pi} u(\rho e^{i\theta})h(e^{i\theta}) \, d\theta \right|$$

$$\le \|h\|_p M_q(\rho, u) \le A_p \|g\|_p M_q(\rho, u).$$

Now take the supremum of $|(1/2\pi)\int vg|$ over all g with $\|g\|_p \le 1$, and use the fact that trigonometric polynomials are dense in L^p. The result is

$$M_q(\rho, v) \le A_p M_q(\rho, u).$$

Thus we may assume $1 < p \le 2$. The next step is to apply Green's theorem in the form

$$r \int_0^{2\pi} \frac{\partial \varphi}{\partial r} d\theta = \iint\limits_{|z| \le r} \nabla^2 \varphi \, dx \, dy,$$

where ∇^2 denotes the Laplacian.

We suppose for the moment that $u(z) > 0$, and compute

$$\nabla^2\{u^p\} = p(p-1)|f'|^2 u^{p-2};$$
$$\nabla^2\{|f|^p\} = p^2 |f'|^2 |f|^{p-2}.$$

Hence

$$\nabla^2\{|f|^p\} \le \frac{p}{p-1} \nabla^2\{u^p\}.$$

By Green's theorem, then,

$$\frac{d}{dr} \int_0^{2\pi} |f(re^{i\theta})|^p \, d\theta \le \frac{p}{p-1} \frac{d}{dr} \int_0^{2\pi} |u(re^{i\theta})|^p \, d\theta.$$

Integrating from 0 to r, and recalling that $v(0) = 0$, we find

$$\int_0^{2\pi} |f(re^{i\theta})|^p \, d\theta - 2\pi[u(0)]^p \le \frac{p}{p-1}\left\{\int_0^{2\pi} |u(re^{i\theta})|^p \, d\theta - 2\pi[u(0)]^p\right\}.$$

Thus

$$M_p(r, v) \le M_p(r, f) \le \left(\frac{p}{p-1}\right)^{1/p} M_p(r, u)$$

if $u(z) > 0$.

For general $u \in h^p$, fix $r < 1$ and set

$$U_1(\theta) = \max\{u(re^{i\theta}), 0\}; \qquad U_2(\theta) = \max\{-u(re^{i\theta}), 0\}.$$

Then

$$u(re^{i\theta}) = U_1(\theta) - U_2(\theta);$$
$$|u(re^{i\theta})|^p = |U_1(\theta)|^p + |U_2(\theta)|^p.$$

$$(4)$$

For $0 \le \rho < 1$, define

$$u_j(\rho e^{i\theta}) = \frac{1}{2\pi} \int_0^{2\pi} P(\rho, \theta - t) U_j(t) \, dt, \qquad j = 1, 2. \tag{5}$$

Let v_j be the harmonic conjugate of u_j. Then

$$u(\rho e^{i\theta}) = u_1(\rho e^{i\theta}) - u_2(\rho e^{i\theta});$$
$$v(\rho e^{i\theta}) = v_1(\rho e^{i\theta}) - v_2(\rho e^{i\theta}).$$

Since u_1 and u_2 are *positive* harmonic functions, it follows from what has been proved that

$$\int_0^{2\pi} |v(\rho re^{i\theta})|^p \, d\theta \le 2^p \left\{ \int_0^{2\pi} |v_1(\rho e^{i\theta})|^p \, d\theta + \int_0^{2\pi} |v_2(\rho e^{i\theta})|^p \, d\theta \right\}$$

$$\le \frac{2^p p}{p-1} \left\{ \int_0^{2\pi} |u_1(\rho e^{i\theta})|^p \, d\theta + \int_0^{2\pi} |u_2(\rho e^{i\theta})|^p \, d\theta \right\}.$$

Letting $\rho \to 1$ and using relations (4) and (5), one now obtains (3) with

$$A_p = 2 \left(\frac{p}{p-1} \right)^{1/p}.$$

The factor 2 can be removed by a more refined argument (see Exercise 5).

As pointed out in Section 3.1, Riesz's theorem breaks down for $p = 1$. The Poisson kernel $P(r, \theta) \in h^1$, but its analytic completion $(1 + z)/(1 - z) \notin H^1$. The next two sections will be concerned with weaker theorems which may be said to replace Riesz's theorem in the case $p = 1$.

Let us note that Riesz's theorem also fails in the case $p = \infty$. Consider, for example, the function

$$f(z) = u(z) + iv(z) = i \log \frac{1 + z}{1 - z},$$

which maps $|z| < 1$ conformally onto the vertical strip $-\pi/2 < u < \pi/2$. Here u is bounded, but v is not. On the other hand, Riesz's theorem guarantees that $f \in H^p$ for all $p < \infty$, a fact not easy to verify by direct calculation.

4.2. KOLMOGOROV'S THEOREM

Although the harmonic conjugate of an h^1 function $u(z)$ need not be in h^1, it does belong to h^p for all $p < 1$. We have already proved this (Corollary to Theorem 3.2) by appeal to Littlewood's subordination theorem. We now give an independent proof of a slightly stronger result. The following lemma will be needed.

LEMMA. For arbitrary positive numbers a and b,

$$(a + b)^p \leq \begin{cases} a^p + b^p, & 0 < p < 1 \\ 2^{p-1}(a^p + b^p), & p > 1. \end{cases}$$

PROOF. It is enough to find the supremum of

$$g(x) = \frac{(1 + x)^p}{1 + x^p}$$

for $x \geq 1$. Differentiation gives

$$g'(x) = p(1 + x)^{p-1}(1 + x^p)^{-2}(1 - x^{p-1}).$$

Thus $g(x)$ is decreasing in the interval $1 \leq x < \infty$ if $p > 1$, increasing if $0 < p < 1$. Since $g(1) = 2^{p-1}$ and $g(x) \to 1$ as $x \to \infty$, the lemma is proved. For $p > 1$, the result also follows easily from the fact that x^p is a convex function.

THEOREM 4.2 (Kolmogorov). If $u \in h^1$, then its conjugate $v \in h^p$ for all $p < 1$. Furthermore, there is a constant B_p, depending only on p, such that

$$M_p(r, v) \leq B_p M_1(r, u), \qquad 0 \leq r < 1,$$

for all $u \in h^1$.

PROOF. Assume first $u(z) > 0$, and set

$$f(z) = u(z) + iv(z) = Re^{i\Phi}, \qquad |\Phi| < \pi/2.$$

Then $F(z) = [f(z)]^p = R^p e^{ip\Phi}$ is analytic in $|z| < 1$. By the mean value theorem,

$$\frac{1}{2\pi} \int_0^{2\pi} F(re^{i\theta}) \, d\theta = F(0) = [u(0)]^p = [M_1(r, u)]^p.$$

Hence

$$\frac{1}{2\pi} \int_0^{2\pi} R^p \cos p\Phi \, d\theta = [M_1(r, u)]^p.$$

In view of the inequalities $|v(z)| < R$ and $0 < \cos p\pi/2 < \cos p\Phi$, it follows that

$$M_p(r, v) < (\sec p\pi/2)^{1/p} M_1(r, u)$$

for *positive u*. The constant is in fact best possible for $u(z) > 0$ (see Exercise 3).

The extension to general $u \in h^1$ is similar to that in the proof of Riesz's theorem. Using the same notation, we find by the lemma that

$$\frac{1}{2\pi} \int_0^{2\pi} |v(\rho r e^{i\theta})|^p \, d\theta \leq \frac{1}{2\pi} \int_0^{2\pi} |v_1(\rho e^{i\theta})|^p \, d\theta + \frac{1}{2\pi} \int_0^{2\pi} |v_2(\rho e^{i\theta})|^p \, d\theta$$

$$\leq \sec p\pi/2\{[M_1(\rho, u_1)]^p + [M_1(\rho, u_2)]^p\}.$$

The other part of the lemma now gives (since $1/p > 1$)

$$M_p(\rho r, v) \leq B_p[M_1(\rho, u_1) + M_1(\rho, u_2)],$$

where

$$B_p = 2^{(1/p)-1}(\sec p\pi/2)^{1/p}.$$

The proof is completed by letting $\rho \to 1$.

4.3. ZYGMUND'S THEOREM

Since $u \in h^1$ is not enough to ensure $v \in h^1$, but the stronger hypothesis $u \in h^p$ (for some $p > 1$) is sufficient, it is natural to ask for the "minimal" growth restriction on u which will imply $v \in h^1$. Such a condition is the boundedness of

$$\int_0^{2\pi} |u(re^{i\theta})| \log^+ |u(re^{i\theta})| \, d\theta, \qquad r < 1.$$

We shall denote by $h \log^+ h$ the class of harmonic functions $u(z)$ for which these integrals are bounded. Clearly, $h^p \subset h \log^+ h$ for all $p > 1$.

THEOREM 4.3 (Zygmund). If $u \in h \log^+ h$, then its conjugate v is of class h^1, and

$$\int_0^{2\pi} |v(re^{i\theta})| \, d\theta \leq \int_0^{2\pi} |u(re^{i\theta})| \log^+ |u(re^{i\theta})| \, d\theta + 6\pi e, \qquad 0 \leq r < 1.$$

PROOF. The proof is similar to that of Riesz's theorem. Assuming first $u(z) \geq e$, one finds

$$\nabla^2\{|f|\} = |f'|^2/|f|, \qquad \nabla^2\{u \log u\} = |f'|^2/u.$$

Thus $\nabla^2\{|f|\} \leq \nabla^2\{u \log u\}$, and an application of Green's theorem gives, as before,

$$\int_0^{2\pi} |f(re^{i\theta})| \, d\theta - 2\pi u(0) \leq \int_0^{2\pi} u(re^{i\theta}) \log u(re^{i\theta}) \, d\theta - 2\pi u(0) \log u(0).$$

From this we conclude [since $u(0) \geq e$]

$$\int_0^{2\pi} |v(re^{i\theta})| \, d\theta \leq \int_0^{2\pi} |u(re^{i\theta})| \log^+ |u(re^{i\theta})| \, d\theta,$$

provided $u(z) \geq e$.

For general $u(z)$, fix $r < 1$ and let

$$U_1(\theta) = \max\{u(re^{i\theta}), e\}, \qquad U_2(\theta) = \max\{-u(re^{i\theta}), e\},$$
$$U_3(\theta) = u(re^{i\theta}) - [U_1(\theta) - U_2(\theta)].$$

Thus $|U_3(\theta)| \leq e$. For $0 \leq \rho < 1$, let

$$u_j(\rho e^{i\theta}) = \frac{1}{2\pi} \int_0^{2\pi} P(\rho, \theta - t) U_j(t) \, dt, \qquad j = 1, 2, 3,$$

and let v_j be the respective harmonic conjugates. Then

$$u(\rho re^{i\theta}) = u_1(\rho e^{i\theta}) - u_2(\rho e^{i\theta}) + u_3(\rho e^{i\theta}),$$
$$v(\rho re^{i\theta}) = v_1(\rho e^{i\theta}) - v_2(\rho e^{i\theta}) + v_3(\rho e^{i\theta}).$$

By what we have already proved,

$$M_1(\rho r, v) \leq M_1(\rho, v_1) + M_1(\rho, v_2) + M_1(\rho, v_3)$$
$$\leq \frac{1}{2\pi} \sum_{j=1}^{2} \int_0^{2\pi} |u_j(\rho e^{i\theta})| \log^+ |u_j(\rho e^{i\theta})| \, d\theta + M_1(\rho, v_3).$$

On the other hand, applying the Schwarz inequality and (2), we find

$$M_1(\rho, v_3) \leq M_2(\rho, v_3) \leq M_2(\rho, u_3) \leq e.$$

Now let ρ tend to 1, with the result

$$2\pi M_1(r, v) \leq \sum_{j=1}^{2} \int_0^{2\pi} |U_j(\theta)| \log^+ |U_j(\theta)| \, d\theta + 2\pi e.$$

To complete the proof, consider the subsets of $[0, 2\pi]$

$$E_1 = \{\theta : u(re^{i\theta}) \geq e\} \quad \text{and} \quad E_2 = \{\theta : u(re^{i\theta}) \leq -e\}.$$

Then

$$\sum_{j=1}^{2} \int_0^{2\pi} |U_j(\theta)| \log^+ |U_j(\theta)| \, d\theta \leq \sum_{j=1}^{2} \int_{E_j} |U_j(\theta)| \log^+ |U_j(\theta)| \, d\theta + 4\pi e$$
$$\leq \int_0^{2\pi} |u(re^{i\theta})| \log^+ |u(re^{i\theta})| \, d\theta + 4\pi e.$$

Zygmund's theorem is, in a sense, best possible: the growth restriction it imposes on $u(z)$ cannot be weakened. The following is a partial converse.

THEOREM 4.4. If both u and its conjugate v belong to h^1, and if $u(z) > C$ for some constant C, then $u \in h \log^+ h$.

PROOF. We may assume $u(z) > 1$, since addition of a constant to u does not affect v. Set

$$f(z) = u(z) + iv(z) = Re^{i\Phi}, \qquad |\Phi| < \pi/2.$$

The function $f(z) \log f(z)$ is analytic in $|z| < 1$. Applying the mean value theorem to its real part, we find

$$\int_0^{2\pi} R \cos \Phi \log R \, d\theta = \int_0^{2\pi} \Phi R \sin \Phi \, d\theta + 2\pi u(0) \log u(0).$$

Thus

$$\int_0^{2\pi} u(re^{i\theta}) \log u(re^{i\theta}) \, d\theta \leq \int_0^{2\pi} u \log R \, d\theta$$

$$< \frac{\pi}{2} \int_0^{2\pi} |v(re^{i\theta})| \, d\theta + 2\pi u(0) \log u(0).$$

COROLLARY. Let $\varphi(x) \geq 0$ be a convex nondecreasing function of $x \geq 0$, such that $\varphi(x) = o(x \log x)$ as $x \to \infty$. Then there exists a harmonic function $u(z)$, with conjugate $v(z)$, such that

$$\int_0^{2\pi} \varphi(|u(re^{i\theta})|) \, d\theta$$

is bounded for $r < 1$, while $v \notin h^1$.

PROOF. Choose an integrable function $U(t) \geq 0$ on $[0, 2\pi]$ such that $\varphi(U(t))$ is integrable but $U(t) \log^+ U(t)$ is not. (We will show in a moment that this is possible.) Let $u(z)$ be the Poisson integral of $U(t)$. Then $u(z) > 0$, and an application of Jensen's inequality (Chapter 2, Exercise 4) shows that $\int \varphi(u(re^{i\theta})) \, d\theta$ is bounded. By the preceding theorem, $v \in h^1$ would imply $u \in h \log^+ h$, which is not the case (by Fatou's lemma). Thus $v \notin h^1$.

It remains to show that a function $U(t)$ can be constructed with the required properties. Suppose, more generally, that $\varphi(x)$ and $\psi(x)$ are any continuous, nonnegative, nondecreasing functions (convex or not), defined on $0 \leq x < \infty$, such that $\psi(x) \to \infty$ and $\varphi(x)/\psi(x) \to 0$ as $x \to \infty$. Choose increasing numbers

$x_n(n = 1, 2, \ldots)$ such that $\psi(x_n) = 2^n$. Then $2^{-n}\varphi(x_n) \to 0$; select a subsequence $\{n_k\}$ for which

$$\sum_{k=1}^{\infty} 2^{-n_k}\varphi(x_{n_k}) < \infty.$$

Now define $t_0 = 0$ and $t_k = t_{k-1} + 2^{-n_k}$, $k \geq 1$. Notice that $0 < t_1 < t_2 < \cdots \leq 1$. Let $T = \lim t_k$. Finally, let $U(t)$ be the step function with the value x_{n_k} in the interval $t_{k-1} \leq t < t_k$ and with $U(t) = 0$ in $T \leq t \leq 2\pi$. Observe that

$$\int_0^{2\pi} \varphi(U(t))\, dt = \sum_{k=1}^{\infty} (t_k - t_{k-1})\varphi(x_{n_k}) + (2\pi - T)\varphi(0)$$

$$= \sum_{k=1}^{\infty} 2^{-n_k}\varphi(x_{n_k}) + (2\pi - T)\varphi(0) < \infty.$$

On the other hand, $\psi(U(t))$ is not integrable because $2^{-n_k}\psi(x_{n_k}) \equiv 1$.

4.4. TRIGONOMETRIC SERIES

The preceding results can also be expressed in the language of trigonometric series. We shall now briefly indicate the connection between the two theories.
A formal trigonometric series

$$\frac{a_0}{2} + \sum_{n=1}^{\infty} (a_n \cos n\theta + b_n \sin n\theta) \tag{6}$$

is called a *Fourier series* if there exists an integrable (real-valued) function $\varphi(t)$ such that

$$a_n = \frac{1}{\pi} \int_0^{2\pi} \cos nt\, \varphi(t)\, dt, \qquad n = 0, 1, 2, \ldots;$$

$$b_n = \frac{1}{\pi} \int_0^{2\pi} \sin nt\, \varphi(t)\, dt, \qquad n = 1, 2, \ldots. \tag{7}$$

It is called a *Fourier–Stieltjes series* if there is a function $\mu(t)$ of bounded variation such that (7) holds with $\varphi(t)\, dt$ replaced by $d\mu(t)$. The formal series

$$\sum_{n=1}^{\infty} (-b_n \cos n\theta + a_n \sin n\theta) \tag{8}$$

is called the *conjugate trigonometric series* of (6).
The use of the word *conjugate* is easily justified. Assuming that (6) is the Fourier series of $\varphi(\theta)$, consider the Poisson integral

$$u(z) = \frac{1}{2\pi} \int_0^{2\pi} P(r, \theta - t)\varphi(t)\, dt.$$

Let

$$f(z) = u(z) + iv(z) = \frac{1}{2\pi} \int_0^{2\pi} \frac{e^{it} + z}{e^{it} - z} \varphi(t) \, dt \qquad (9)$$

be its analytic completion, normalized as usual by $v(0) = 0$. The expansion

$$\frac{e^{it} + z}{e^{it} - z} = 1 + 2 \sum_{n=1}^{\infty} e^{-int} z^n$$

and the relations (7) now give

$$f(z) = \frac{a_0}{2} + \sum_{n=1}^{\infty} c_n z^n, \qquad c_n = a_n - ib_n.$$

Taking real and imaginary parts, we find

$$u(z) = \frac{a_0}{2} + \sum_{n=1}^{\infty} r^n (a_n \cos n\theta + b_n \sin n\theta),$$

$$v(z) = \sum_{n=1}^{\infty} r^n (-b_n \cos n\theta + a_n \sin n\theta).$$

These series are convergent for every $r < 1$. On the other hand

$$u(re^{i\theta}) \to \varphi(\theta) \quad \text{a.e.,}$$

by Theorem 1.2. In the language of summability theory, we may therefore say that the series (6) is *Abel summable* to $\varphi(\theta)$ for almost every θ. It is known, however, that if φ is merely integrable, the series (6) need not converge anywhere in the usual sense.

A similar statement may be made about the conjugate series (8). If (6) is a Fourier series, or even a Fourier–Stieltjes series, then $u \in h^1$. Hence, by Kolmogorov's theorem, $v \in h^p$ for all $p < 1$, and (by Theorem 2.2) $v(z)$ has a radial limit a.e. The function

$$\tilde{\varphi}(\theta) = \lim_{r \to 1} v(re^{i\theta})$$

is called the *conjugate function* of $\varphi(\theta)$. Thus the conjugate series of the Fourier series of φ is Abel summable almost everywhere to $\tilde{\varphi}(\theta)$, and $\tilde{\varphi} \in L^p$ for all $p < 1$.

Must $\tilde{\varphi}$ be integrable? If so, then $v \in h^1$ (Theorem 2.11), and $v(z)$ is the Poisson integral of $\tilde{\varphi}$; thus (8) is actually the Fourier series of $\tilde{\varphi}$. Conversely, if (8) is the Fourier series of some integrable function, that function must be $\tilde{\varphi}$. Thus $\tilde{\varphi} \in L^1$ if and only if the conjugate series (8) is itself a Fourier series, or if and only if $v(z) \in h^1$. If $\tilde{\varphi}$ is not integrable, then the conjugate series is not even a Fourier–Stieltjes series (by Theorem 1.1). An example constructed in Section 4.5 will show that this can actually happen.

If φ is more than integrable, a correspondingly stronger statement can be made about $\tilde{\varphi}$. Suppose, for example, that $\varphi \in L^p$ for some p, $1 < p < \infty$. Then $u \in h^p$, so $v \in h^p$ by Riesz's theorem, and $\tilde{\varphi} \in L^p$. In fact, there is a constant A_p such that $\|\tilde{\varphi}\|_p \leq A_p \|\varphi\|_p$. If $\varphi \in L \log^+ L$ (notation obvious), then $u \in h \log^+ h$, and Zygmund's theorem tells us that $v \in h^1$; hence $\tilde{\varphi} \in L^1$. It was in terms of φ and $\tilde{\varphi}$ that these theorems were originally formulated.

There is another representation of $\tilde{\varphi}(\theta)$ which is more direct and which is convenient for some purposes. If we take the imaginary part of (9) and let $r \to 1$ under the integral sign, we find formally

$$\tilde{\varphi}(\theta) = \frac{1}{2\pi} \int_0^{2\pi} \cot\left(\frac{\theta - t}{2}\right) \varphi(t)\, dt.$$

Unfortunately, this last integral need not converge, because of the singularity at $t = \theta$. The formula is correct, however, if the integral is interpreted in the Cauchy principal-value sense. That is, after changing variables and extending φ periodically,

$$\tilde{\varphi}(\theta) = \lim_{\varepsilon \to 0} \frac{1}{2\pi} \int_\varepsilon^\pi \cot(t/2)[\varphi(\theta - t) - \varphi(\theta + t)]\, dt \quad \text{a.e.}$$

We omit the proof, since we shall make no use of this formula. Proofs may be found in most books on trigonometric series; for example, in Bary [1], Vol. 2, pp. 60–62.

4.5. THE CONJUGATE OF AN h^1 FUNCTION

If $u(z)$ and its conjugate $v(z)$ both belong to h^1, then $u(z)$ is a Poisson integral (Theorem 3.1); that is, the function $\mu(t)$ in its Poisson–Stieltjes representation must be absolutely continuous. The converse is false. Even if $u(z)$ is a Poisson integral, $v(z)$ need not be of class h^1. We are about to present a counter-example.

In view of Hardy's inequality (Corollary to Theorem 3.15), the function

$$f(z) = \sum_{n=2}^\infty \frac{z^n}{\log n} \notin H^1.$$

However, its real part belongs to h^1, and is in fact a Poisson integral. To see this, it is enough to observe that

$$\sum_{n=2}^\infty \frac{\cos nt}{\log n}$$

is the Fourier series of an integrable function. But this is a particular case of the following more general proposition.

THEOREM 4.5. If the real sequence $\{a_0, a_1, \ldots\}$ is convex and $a_n \to 0$ as $n \to \infty$, then the series

$$\frac{a_0}{2} + \sum_{n=1}^{\infty} a_n \cos nt \tag{10}$$

converges for all $t \neq 0$ in $[-\pi, \pi]$ to a nonnegative integrable function $\varphi(t)$, and (10) is the Fourier series of $\varphi(t)$.

PROOF. Let $\Delta a_n = a_{n+1} - a_n$ and $\Delta^2 a_n = \Delta(\Delta a_n)$. The convexity of $\{a_n\}$ means that $\Delta^2 a_n \geq 0$; thus $\{\Delta a_n\}$ is an increasing sequence. Since $\{a_n\}$ converges, $\Delta a_n \to 0$; hence $\Delta a_n \leq 0$. Since $a_n \to 0$, it follows that $a_n \geq 0$.

Let $S_N(t)$ denote the Nth partial sum of (10). After two summations by parts, we find

$$S_N(t) = \sum_{n=0}^{N} (\Delta^2 a_n)(n + 1)K_n(t)$$

$$-(\Delta a_{N+1})(N + 1)K_N(t) + a_{N+1}D_N(t), \tag{11}$$

where

$$D_N(t) = \frac{1}{2} + \sum_{n=1}^{N} \cos nt = \frac{\sin(N + \frac{1}{2})t}{2 \sin t/2}$$

and

$$K_N(t) = \frac{1}{N+1} \sum_{n=0}^{N} D_n(t) = \frac{\sin^2 (N + 1)t/2}{2(N + 1) \sin^2 (t/2)}$$

are the Dirichlet and the Fejér kernel, respectively. For each $t \neq 0$, the sequences $\{D_N(t)\}$ and $\{NK_{N-1}(t)\}$ are bounded; hence the last two terms in (11) tend to zero. Consequently,

$$S_N(t) \to \varphi(t) = \sum_{n=0}^{\infty} (\Delta^2 a_n)(n + 1)K_n(t). \tag{12}$$

Because $(n + 1)K_n(t)$ is uniformly bounded in each set $0 < \delta \leq |t| \leq \pi$, the series (12) converges uniformly there, and so represents a function $\varphi(t)$ continuous for $t \neq 0$. Since the terms of the series are all nonnegative, so is $\varphi(t)$. By the Lebesgue monotone convergence theorem,

$$\int_{-\pi}^{\pi} \varphi(t) \cos mt \, dt = \sum_{n=0}^{\infty} (\Delta^2 a_n)(n + 1) \int_{-\pi}^{\pi} K_n(t) \cos mt \, dt$$

$$= \pi \sum_{n=m}^{\infty} (\Delta^2 a_n)(n - m + 1), \qquad m = 0, 1, 2, \ldots.$$

This last series may be evaluated through summation by parts:

$$\sum_{n=m}^{k} (\Delta^2 a_n)(n - m + 1) = -\sum_{n=m}^{k} \Delta a_n + (k - m + 1) \Delta a_{k+1}$$

$$= a_m - a_{k+1} + (k - m + 1) \Delta a_{k+1}.$$

As $k \to \infty$, the right-hand side approaches a_m, and so

$$\frac{1}{\pi} \int_{-\pi}^{\pi} \varphi(t) \cos mt \, dt = a_m, \qquad m = 0, 1, 2, \ldots.$$

[In particular, $\varphi(t)$ is integrable.] Thus (10) is a Fourier series, which was to be proved.

The fact that $k \, \Delta \, a_k \to 0$ follows from a classical theorem of Abel: *If $\{b_k\}$ is a monotonically decreasing sequence and $\sum b_k$ converges, then $kb_k \to 0$.* For a proof, observe that

$$2^n b_{2^n} \leq 2 \sum_{k=2^n-1}^{2^n} b_k.$$

4.6. THE CASE $p < 1$: A COUNTEREXAMPLE

For $1 < p < \infty$, the class h^p is preserved under conjugation. This is false for h^1, but it is "almost true" in the sense that the conjugates of h^1 functions belong to h^p for all $p < 1$. If the hypothesis is further weakened by requiring only that $u \in h^p$ for some $p < 1$, hardly a trace of the theorem remains. The conjugate function v may not belong to h^q for any positive q. In fact, we are about to construct an analytic function $f(z) = u(z) + iv(z)$ such that $u \in h^p$ for all $p < 1$, yet $v \notin h^p$ for any $p > 0$.

The example is

$$f(z) = u(z) + iv(z) = \sum_{n=1}^{\infty} \varepsilon_n \frac{z^{2^n}}{1 - z^{2^{n+1}}}, \qquad \varepsilon_n = \pm 1. \tag{13}$$

We are going to show that for every choice of the signs ε_n, $u \in h^p$ for all $p < 1$; while for "almost every" sequence of signs, $f(z)$ has a radial limit on no set of positive measure. In particular, *some* choice of the ε_n gives a function $f(z)$ which is not even of class N, but whose real part belongs to h^p for all $p < 1$.

LEMMA. For each $p > \frac{1}{2}$,

$$\int_{-\pi}^{\pi} \frac{d\theta}{(1 - 2r \cos \theta + r^2)^p} = O\left(\frac{1}{(1 - r)^{2p-1}}\right) \qquad \text{as} \quad r \to 1.$$

PROOF. Since $\sin x \geq (2/\pi)x$ for $0 \leq x \leq \pi/2$,

$$1 - 2r \cos \theta + r^2 = (1 - r)^2 + 4r \sin^2 \theta/2 \geq (1 - r)^2 + (4r/\pi^2)\theta^2.$$

Hence, for $r \geq \frac{1}{2}$, the integral does not exceed

$$\int_{-\pi}^{\pi} \frac{d\theta}{[(1 - r)^2 + 2\pi^{-2}\theta^2]^p} < \frac{1}{(1 - r)^{2p-1}} \int_{-\infty}^{\infty} \frac{dt}{[1 + 2\pi^{-2}t^2]^p}.$$

The last integral is convergent because $p > \frac{1}{2}$.

In showing $u \in h^p$ for all $p < 1$, we may suppose $p > \frac{1}{2}$. With $z = re^{i\theta}$, we begin by computing

$$\operatorname{Re}\left\{\frac{z^{2^n}}{1 - z^{2^{n+1}}}\right\} = \frac{R_n(1 - R_n^2) \cos 2^n\theta}{1 - 2R_n^2 \cos 2^{n+1}\theta + R_n^4}, \qquad R_n = r^{2^n}.$$

From this we find, using the lemma,

$$\int_0^{2\pi} |u(re^{i\theta})|^p \, d\theta \leq \sum_{n=1}^{\infty} R_n^p(1 - R_n^2)^p \int_0^{2\pi} \frac{d\theta}{(1 - 2R_n^2 \cos \theta + R_n^4)^p}$$

$$\leq A \sum_{n=1}^{\infty} R_n^p(1 - R_n^2)^{1-p} \leq 2^{1-p}A \sum_{n=1}^{\infty} R_n^p(1 - R_n)^{1-p},$$

where A is a constant. But

$$R_n^p(1 - R_n)^{1-p} < r^{2^np}2^{n(1-p)}(1 - r)^{1-p},$$

since $1 - r^m = (1 - r)(1 + r + \cdots + r^{m-1}) < m(1 - r)$. Furthermore,

$$\frac{1 - r^p}{1 - r} = p + O(1 - r) \qquad (r \to 1),$$

by the binomial expansion of $r^p = [1 - (1 - r)]^p$. We will have shown $u \in h^p$, then, if we prove

$$\sum_{n=1}^{\infty} 2^{n\alpha}r^{2^n} = O((1 - r)^{-\alpha}), \qquad \alpha > 0. \tag{14}$$

It is equivalent to estimate the integral

$$\int_0^{\infty} 2^{x\alpha}r^{2^x} \, dx = \frac{1}{\log 2} \int_1^{\infty} e^{u \log r}u^{\alpha-1} \, du \qquad (u = 2^x).$$

Since $\log r < r - 1$, the second integral is majorized by

$$\int_1^{\infty} e^{-(1-r)u}u^{\alpha-1} \, du = (1 - r)^{-\alpha} \int_{1-r}^{\infty} e^{-v}v^{\alpha-1} \, dv,$$

which proves (14). Thus $u \in h^p$ for all $p < 1$.

To prove the nonexistence of radial limits we appeal to the theory of Rademacher functions, as developed in Appendix A, and to Theorem A.4 in particular. For the present application we have

$$g_n(z) = \frac{z^{2^n}}{1 - z^{2^{n+1}}}.$$

Thus

$$\sum_{n=N}^{\infty} |g_n(z)|^2 = \sum_{n=N}^{\infty} R_n^2 (1 - 2R_n^2 \cos 2^{n+1}\theta + R_n^4)^{-1}$$

$$\geq \frac{1}{4} \sum_{n=N+1}^{\infty} r^{2^n} \to \infty$$

as $r \to 1$. Since the other hypotheses of Theorem A.4 are obviously satisfied, we conclude that for almost every choice of signs, the analytic function (13) has a radial limit almost nowhere, which is what we wanted to prove.

EXERCISES

1. Show that if $\varphi \in L^p$ $(1 < p < \infty)$ and

$$\varphi(t) \sim \sum_{n=-\infty}^{\infty} a_n e^{int}$$

then its "analytic projection" $f(z) = \sum_{n=0}^{\infty} a_n z^n$ belongs to H^p, and $\|f\|_p \leq A_p \|\varphi\|_p$ for some constant A_p independent of φ.

2. Show that the Poisson kernel $P(r, \theta)$ does not belong to h^p for any $p > 1$.

3. Show that the constant $B_p = (\sec p\pi/2)^{1/p}$ is best possible for the special case of Kolmogorov's theorem in which $u(z) > 0$. (*Hint*: Try the Poisson kernel.)

4. Use Green's theorem to prove Kolmogorov's theorem with $B_p = 2^{1/p-1}(1-p)^{-1/p}$.

5. Let $f(z) = u(z) + iv(z)$ be analytic in $|z| < 1$, and suppose $u \in h^p$ for some p, $1 < p \leq 2$. For fixed $\varepsilon > 0$, set

$$G(z) = \{[u(z)]^2 + \varepsilon\}^{p/2}; \qquad H(z) = \{|f(z)|^2 + \varepsilon\}^{p/2}.$$

Show that

$$\nabla^2 H \leq \frac{p}{p-1} \nabla^2 G.$$

Apply Green's theorem, integrate, and let $\varepsilon \to 0$ to obtain

$$M_p(r, f) \leq \left(\frac{p}{p-1}\right)^{1/p} M_p(r, u),$$

thus proving the theorem of M. Riesz with a smaller constant. (This idea is due to W. K. Hayman. It is an open problem to find the best possible constant, even for $u(z) > 0$. The above constant reduces to $\sqrt{2}$ for $p = 2$, while the best possible constant in this case is 1.)

6. For the example of Section 4.6, show that there is a choice of signs such that

$$M_p(r, f) \neq O\left(\left(\log \frac{1}{1-r}\right)^{1/2}\right).$$

NOTES

Theorem 4.1, or an equivalent form of it, is in the paper of M. Riesz [1]. The proof based on Green's theorem, as presented in the text, is due to P. Stein [1]. This approach has the advantage of leading to a relatively good value of the constant A_p. Calderón [1] has given still another proof; see also Zygmund [4], Chap. VII. Shortly after Kolmogorov [1] proved Theorem 4.2, Littlewood [2] suggested the proof using his subordination theorem. Hardy [3] then discovered the elementary argument given in the text as a second proof. Theorem 4.3 and the converse results (Theorem 4.4 and its corollary) are due to Zygmund [1, 4]. Another proof is in a paper of Littlewood [3]. Theorem 4.5 and similar results may be found in Zygmund [4], Chap. V. Phenomena of the type illustrated in Sec. 4.6 are discussed in the paper of Hardy and Littlewood [6]. Previously, Littlewood [1] had constructed a harmonic function which belongs to h^p for all $p < 1$, yet has a radial limit almost nowhere. The example (13) given in Sec. 4.6 is essentially due to Hardy and Littlewood [6], who showed by a highly nonelementary argument that a function similar to this (but with $\varepsilon_n \equiv 1$) fails to have a radial limit on some set of positive measure. Paley and Zygmund [2] introduced Rademacher functions and "constructed" an example of the type (13) for which $f \notin H^p$ for all $p < 1$. The argument given in the text is considerably simpler than theirs. Hardy and Littlewood [6] also proved that if $u \in h^p$ for some $p \leq 1$, then its conjugate v satisfies

$$M_p(r, v) = O\left(\left(\log \frac{1}{1-r}\right)^{1/p}\right), \qquad r \to 1.$$

For $p = 1, \frac{1}{2}, \frac{1}{3}, \ldots$, they showed by an elementary example that this estimate is best possible. Whether or not it can be improved for other values of p

remains an open question, although Swinnerton-Dyer [1] has shown that it cannot be improved to

$$M_p(r, v) = o\left(\log \frac{1}{1 - r}\right).$$

(Unfortunately, Hardy and Littlewood stated the positive result incorrectly in the introduction to their paper, and Swinnerton-Dyer reproduced this error.) Gwilliam [1] simplified some of the work of Hardy and Littlewood in this area.

If $f(z)$ is analytic in the unit disk, there is a very close relation between the mean growth of the derivative $f'(z)$ and the smoothness of the boundary function $f(e^{i\theta})$. This principle takes several precise forms, as we shall see in the present chapter. Other topics to be discussed are the relations between the growth of $M_p(r, f)$ and $M_p(r, f')$, between the growth of $M_p(r, f)$ and $M_q(r, f)$, and between the growth and smoothness of a harmonic function and its conjugate. Some of these results will be applied repeatedly in later chapters. We shall begin by discussing several ways to measure smoothness and exploring the connections between them.

5.1. SMOOTHNESS CLASSES

Let $\varphi(x)$ be a complex-valued function defined on $-\infty < x < \infty$ and periodic with period 2π. The *modulus of continuity* of φ is the function

$$\omega(t) = \omega(t; \varphi) = \sup_{|x-y| \le t} |\varphi(x) - \varphi(y)|.$$

Thus φ is continuous if and only if $\omega(t) \to 0$ as $t \to 0$. We say that φ is of class

Λ_α $(0 < \alpha \le 1)$ if $\omega(t) = O(t^\alpha)$ as $t \to 0$. Alternatively, Λ_α is the class of functions which satisfy a *Lipschitz condition* of order α:

$$|\varphi(x) - \varphi(y)| \le A\,|(x - y)|^\alpha.$$

The definition is of no interest for $\alpha > 1$, since Λ_α would then contain only the constant functions. It is clear that $\Lambda_\beta \subset \Lambda_\alpha$ if $\alpha < \beta$.

A continuous function $\varphi(x)$ is said to be of class Λ_* if there is a constant A such that

$$|\varphi(x + h) - 2\varphi(x) + \varphi(x - h)| \le Ah \tag{1}$$

for all x and for all $h > 0$. The condition (1) alone does not imply continuity; indeed, it is well known that there exist nonmeasurable functions such that

$$\varphi(x + y) = \varphi(x) + \varphi(y).$$

For any $\alpha < 1$, the proper inclusions

$$\Lambda_1 \subset \Lambda_* \subset \Lambda_\alpha$$

hold. In fact, every $\varphi \in \Lambda_*$ has modulus of continuity $\omega(t) = O(t \log 1/t)$, as we shall see later. (See also Exercise 14.) It often turns out that Λ_*, not Λ_1, is the "natural limit" of Λ_α as $\alpha \to 1$. Another class, apparently larger than Λ_α, could be defined by requiring that the left-hand side of (1) be $O(h^\alpha)$. However, this class is actually the same as Λ_α if $\alpha < 1$.

For $\varphi \in L^p = L^p(0, 2\pi)$, $1 \le p < \infty$, the function

$$\omega_p(t) = \omega_p(t; \varphi) = \sup_{0 < h \le t} \left\{ \int_0^{2\pi} |\varphi(x + h) - \varphi(x)|^p dx \right\}^{1/p}$$

is called the *integral modulus of continuity* of order p. For every L^p function φ, $\omega_p(t) \to 0$ as $t \to 0$, by a theorem of F. Riesz (see, for example, Titchmarsh [1], p. 397). If $\omega_p(t) = O(t^\alpha)$, $0 < \alpha \le 1$, we say that φ belongs to the class $\Lambda_\alpha{}^p$. Because the L^p means increase with p, it is evident that $\Lambda_\alpha{}^q \subset \Lambda_\alpha{}^p$ if $p < q$. If φ is continuous, $\omega_p(t) \to \omega(t)$ as $p \to \infty$. The classes $\Lambda_\alpha{}^p$ may therefore be viewed as generalizations of Λ_α. Obviously, $\Lambda_\alpha \subset \Lambda_\alpha{}^p$.

In the proof of Theorem 5.4 we shall need the following result, which is of some interest in itself.

LEMMA 1 (Hardy–Littlewood). If φ is of bounded variation over $[0, 2\pi]$, then $\varphi \in \Lambda_1{}^1$. Conversely, every function $\varphi \in \Lambda_1{}^1$ is equal almost everywhere to a function of bounded variation.

PROOF. Suppose first that $\varphi(x)$ is of bounded variation, and let $V(x)$ be the total variation of φ on $[0, x]$. Then for small $h > 0$,

$$\int_0^{2\pi} |\varphi(x+h) - \varphi(x)| \, dx \le \int_0^{2\pi} [V(x+h) - V(x)] \, dx$$

$$= \int_{2\pi}^{2\pi+h} V(x) \, dx - \int_0^h V(x) \, dx \le Ch.$$

The converse is more difficult. Assuming $\varphi \in \Lambda_1{}^1$, let

$$F(x) = \int_0^x \varphi(t) \, dt$$

and

$$\varphi_n(x) = n[F(x+1/n) - F(x)].$$

Then $\varphi_n(x)$ is absolutely continuous and $\varphi_n(x) \to \varphi(x)$ a.e. as $n \to \infty$. For $h > 0$ we have

$$\varphi_n(x+h) - \varphi_n(x) = n \int_0^h \left[\varphi\left(x + \frac{1}{n} + t\right) - \varphi(x+t)\right] dt.$$

Since $\varphi \in \Lambda_1{}^1$, it follows that

$$\int_0^{2\pi} |\varphi_n(x+h) - \varphi_n(x)| \, dx \le n \int_0^h dt \int_0^{2\pi} \left|\varphi\left(x + \frac{1}{n} + t\right) - \varphi(x+t)\right| dx$$

$$\le Ch,$$

where the constant C is independent of n. Thus, by Fatou's lemma

$$\int_0^{2\pi} |\varphi_n{}'(x)| \, dx \le \lim_{h \to 0} \inf \frac{1}{h} \int_0^{2\pi} |\varphi_n(x+h) - \varphi_n(x)| \, dx \le C.$$

Now let $0 = x_0 < x_1 < \cdots < x_m = 2\pi$ be an arbitrary finite partition of the interval $[0, 2\pi]$. Bearing in mind that $\varphi_n(x)$ is absolutely continuous, we have

$$\sum_{k=1}^m |\varphi_n(x_k) - \varphi_n(x_{k-1})| \le \int_0^{2\pi} |\varphi_n{}'(x)| \, dx \le C.$$

But $\varphi_n(x) \to \varphi(x)$ if x is not in a certain set E of measure zero. Hence for any partition such that $x_k \notin E$,

$$\sum_{k=1}^m |\varphi(x_k) - \varphi(x_{k-1})| \le C. \tag{2}$$

It can now be shown from (2), by imitating the classical argument (see, for example, Rudin [7]) that $\varphi(x)$ coincides on the complement of E with the difference of two monotonic functions. Consequently, $\varphi(x)$ has a natural extension to $[0, 2\pi]$, and the extended function satisfies (2) for *all* partitions. This proves that $\varphi(x)$ coincides almost everywhere with a function of bounded variation.

5.2. SMOOTHNESS OF THE BOUNDARY FUNCTION

Generally speaking, it is reasonable to expect an analytic function to be smooth on the boundary if its derivative grows slowly, and conversely. For the unit disk, this principle can be expressed in surprisingly precise form, as follows.

THEOREM 5.1 (Hardy–Littlewood). Let $f(z)$ be a function analytic in $|z| < 1$. Then $f(z)$ is continuous in $|z| \le 1$ and $f(e^{i\theta}) \in \Lambda_\alpha$ $(0 < \alpha \le 1)$, if and only if

$$f'(z) = O\left(\frac{1}{(1-r)^{1-\alpha}}\right).$$

PROOF. First let us dispose of the case $\alpha = 1$. By Theorem 3.11, the Lipschitz condition on the boundary implies that $f' \in H^1$ and that $f'(e^{i\theta}) \in L^\infty$. Hence $f' \in H^\infty$, by Theorem 2.11. Conversely, if $f' \in H^\infty$, another application of Theorem 3.11 shows that $f(z)$ is continuous in $|z| \le 1$, $f(e^{i\theta})$ is absolutely continuous, and $(d/d\theta)f(e^{i\theta}) \in L^\infty$. Integration of the derivative therefore gives $f(e^{i\theta}) \in \Lambda_1$.

Now let $f(z)$ be continuous in $|z| \le 1$, and suppose $f(e^{i\theta}) \in \Lambda_\alpha$, $0 < \alpha < 1$. By the Cauchy formula,

$$f'(z) = \frac{1}{2\pi} \int_0^{2\pi} \frac{[f(e^{it}) - f(e^{i\theta})]e^{it}}{(e^{it} - z)^2} \, dt, \qquad z = re^{i\theta}.$$

Thus

$$|f'(z)| \le \frac{1}{2\pi} \int_{-\pi}^{\pi} \frac{|f(e^{i(t+\theta)}) - f(e^{i\theta})|}{1 - 2r\cos t + r^2} \, dt. \tag{3}$$

Using the Lipschitz condition and the relation

$$1 - 2r\cos t + r^2 = (1-r)^2 + 4r\sin^2 \frac{t}{2} \ge (1-r)^2 + \frac{4rt^2}{\pi^2},$$

$0 \le t \le \pi$, we find

$$|f'(z)| \le \frac{A}{2\pi} \int_{-\pi}^{\pi} \frac{|t|^\alpha \, dt}{(1-r)^2 + 4r(t/\pi)^2} = O\left(\frac{1}{(1-r)^{1-\alpha}}\right).$$

The assumption $\alpha < 1$ assures the convergence of the integral

$$\int_0^\infty \frac{u^\alpha \, du}{1 + u^2},$$

which arises after the substitution $u = t/(1-r)$.

Conversely, suppose

$$|f'(z)| \le \frac{C}{(1-r)^{1-\alpha}}, \qquad 0 < \alpha < 1.$$

Then the radial limit

$$f(e^{i\theta}) = f(0) + \lim_{R \to 1} \int_0^R f'(re^{i\theta})\, dr$$

exists for every θ. Furthermore, $f \in H^\infty$, so $f(z)$ is the Poisson integral of $f(e^{i\theta})$. Hence the continuity of $f(e^{i\theta})$ would imply the continuity of $f(z)$ in $|z| \le 1$. We shall prove the continuity by showing $f(e^{i\theta}) \in \Lambda_\alpha$. For this purpose, choose θ and φ with $0 < \varphi - \theta < 1$. Fix ρ, $0 < \rho < 1$, and let Γ be the contour consisting of the radial segment from $e^{i\theta}$ to $\rho e^{i\theta}$, the arc of the circle $|z| = \rho$ from $\rho e^{i\theta}$ to $\rho e^{i\varphi}$, and the radial segment from $\rho e^{i\varphi}$ to $e^{i\varphi}$. (See Fig. 2.) Then

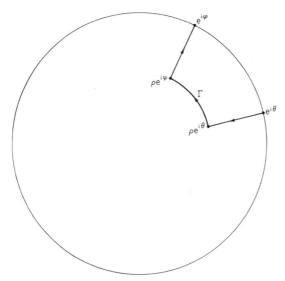

Figure 2

$$f(e^{i\varphi}) - f(e^{i\theta}) = \int_\Gamma f'(\zeta)\, d\zeta.$$

Breaking up the integral into its three components, we find

$$|f(e^{i\varphi}) - f(e^{i\theta})| \le \int_\rho^1 |f'(re^{i\theta})|\, dr + \int_\rho^1 |f'(re^{i\varphi})|\, dr + \int_\theta^\varphi |f'(\rho e^{it})|\, dt$$

$$\le 2C \int_\rho^1 \frac{dr}{(1-r)^{1-\alpha}} + \frac{C(\varphi - \theta)}{(1-\rho)^{1-\alpha}}.$$

With the choice $\rho = 1 - (\varphi - \theta)$, this gives

$$|f(e^{i\varphi}) - f(e^{i\theta})| \le C\left(1 + \frac{2}{\alpha}\right)|\varphi - \theta|^{\alpha},$$

and the proof is complete.

The method may be used to prove other theorems of the same type. We mention one example, leaving the proof as an exercise.

THEOREM 5.2. Let $f(z)$ be analytic in $|z| < 1$, and suppose

$$f'(z) = O\left(\log\frac{1}{1-r}\right).$$

Then $f(z)$ is continuous in $|z| \le 1$, and $f(e^{i\theta})$ has modulus of continuity

$$\omega(t) = O\left(t \log\frac{1}{t}\right).$$

According to Theorem 5.1, $f' \in H^{\infty}$ if and only if $f(e^{i\theta}) \in \Lambda_1$. If the growth condition is slightly weakened to $f''(z) = O((1 - r)^{-1})$, Theorem 5.2 says that $f(e^{i\theta})$ must still be "almost" of class Λ_1: $\omega(t) = O(t \log 1/t)$. This degree of smoothness, however, does not ensure that $f''(z) = O((1 - r)^{-1})$. The necessary and sufficient condition is that the boundary function belong to the class Λ_* introduced in Section 5.1. The precise result is as follows.

THEOREM 5.3 (Zygmund). Let $f(z)$ be analytic in $|z| < 1$. Then $f(z)$ is continuous in $|z| \le 1$ and $f(e^{i\theta}) \in \Lambda_*$ if and only if

$$f''(z) = O\left(\frac{1}{1-r}\right). \tag{4}$$

COROLLARY. If $f(z)$ is analytic in $|z| < 1$, continuous in $|z| \le 1$, and $f(e^{i\theta}) \in \Lambda_*$ then $f(e^{i\theta})$ has modulus of continuity $\omega(t) = O(t \log 1/t)$.

PROOF OF THEOREM. If $f(z)$ is continuous in $|z| \le 1$, it can be represented as a Poisson integral:

$$f(z) = \frac{1}{2\pi} \int_{-\pi}^{\pi} P(r, \theta - t) f(e^{it})\, dt, \qquad z = re^{i\theta}. \tag{5}$$

Since the second derivative $P_{\theta\theta}(r, \theta)$ is an even function of θ and

$$\int_0^{\pi} P_{\theta\theta}(r, t)\, dt = P_{\theta}(r, \pi) - P_{\theta}(r, 0) = 0,$$

it follows that

$$f_{\theta\theta}(z) = \frac{1}{2\pi} \int_0^\pi P_{\theta\theta}(r, t)\{f(e^{i(\theta+t)}) - 2f(e^{i\theta}) + f(e^{i(\theta-t)})\}\, dt.$$

The hypothesis $f(e^{i\theta}) \in \Lambda_*$ therefore implies

$$|f_{\theta\theta}(z)| \le A \int_0^\pi t P_{\theta\theta}(r, t)\, dt = A[P(r, 0) - P(r, \pi)] = O\!\left(\frac{1}{1-r}\right),$$

since $P_{\theta\theta}(r, \theta) \ge 0$ for $0 \le \theta \le \pi$ and $r \ge 2 - \sqrt{3}$, as a calculation shows. On the other hand, (5) and the boundedness of $f(e^{it})$ easily show

$$f_\theta(z) = O\!\left(\frac{1}{1-r}\right).$$

Thus

$$f''(z) = r^{-2}e^{-2i\theta}\{if_\theta(z) - f_{\theta\theta}(z)\} = O\!\left(\frac{1}{1-r}\right).$$

The proof of the converse is more difficult. In view of Theorem 5.1, the condition (4) implies $f(z)$ is continuous in $|z| \le 1$; it must be shown that $f(e^{i\theta}) \in \Lambda_*$. For $0 < h < 1$, let us use the notation

$$\Delta_h\, G(\theta) = G(\theta + h) - G(\theta),$$
$$\Delta_h^2\, G(\theta) = G(\theta + 2h) - 2G(\theta + h) + G(\theta).$$

We are required to show that $\Delta_h^2\, f(e^{i\theta}) = O(h)$, uniformly in θ, as $h \to 0$. Our strategy is to write

$$\Delta_h^2 f(e^{i\theta}) = \Delta_h^2\{f(e^{i\theta}) - f(\rho e^{i\theta})\} + \Delta_h^2 f(\rho e^{i\theta}) \tag{6}$$

$(0 < \rho < 1)$, to set $\rho = 1 - h$, and to show that as $h \to 0$, each of the two terms in (6) is uniformly $O(h)$.

The identity

$$f(e^{i\theta}) - f(\rho e^{i\theta}) = (1 - \rho)e^{i\theta}f'(\rho e^{i\theta}) + e^{2i\theta}\int_\rho^1 (1 - r)f''(re^{i\theta})\, dr \tag{7}$$

is easily verified through integration by parts. Now set $\rho = 1 - h$. Under the hypothesis (4), the integral in (7) is then uniformly $O(h)$. Thus

$$\Delta_h^2\{f(e^{i\theta}) - f(\rho e^{i\theta})\} = h\, \Delta_h^2\{e^{i\theta}f'(\rho e^{i\theta})\} + O(h). \tag{8}$$

But

$$\Delta_h\{e^{i\theta}f'(\rho e^{i\theta})\} = \Delta_h\{e^{i\theta}\}f'(\rho e^{i(\theta+h)}) + e^{i\theta}\, \Delta_h\{f'(\rho e^{i\theta})\}$$

$$= O\!\left(h \log \frac{1}{h}\right) + i\rho e^{i\theta}\int_0^h e^{i(\theta+t)}f''(\rho e^{i(\theta+t)})\, dt = O(1), \tag{9}$$

uniformly in θ. Hence the expression (8) is uniformly $O(h)$.

Finally,

$$\Delta_h^2 f(\rho e^{i(\theta-h)}) = i\rho \int_0^h \{e^{i(\theta+t)}f'(\rho e^{i(\theta+t)}) - e^{i(\theta-t)}f'(\rho e^{i(\theta-t)})\}\, dt.$$

Analyzing the integrand as in (9), we see that it is dominated by

$$Ct\left\{\log\frac{1}{h} + \frac{1}{h}\right\},$$

where C is independent of θ. Hence $\Delta_h^2 f(\rho e^{i\theta})$ is uniformly $O(h)$, and the proof is complete.

The next theorem may be viewed as the L^p analogue of Theorem 5.1. It is included here for the sake of completeness; we shall make no use of it in this book.

THEOREM 5.4 (Hardy–Littlewood). Let $f(z)$ be analytic in $|z| < 1$. Then $f \in H^p$ and $f(e^{i\theta}) \in \Lambda_\alpha^p$ $(1 \le p < \infty; 0 < \alpha \le 1)$ if and only if

$$M_p(r, f') = O\left(\frac{1}{(1-r)^{1-\alpha}}\right). \tag{10}$$

PROOF. We first show that the condition (10) for any $\alpha > 0$ implies $f \in H^p$. It is convenient to assume $f(0) = 0$, so that

$$f(re^{i\theta}) = \int_0^r f'(\rho e^{i\theta})e^{i\theta}\, d\rho.$$

Applying the continuous form of Minkowski's inequality, we find

$$M_p(r, f) \le \int_0^r M_p(\rho, f')\, d\rho \le C\int_0^r \frac{d\rho}{(1-\rho)^{1-\alpha}} < \frac{C}{\alpha};$$

hence $f \in H^p$.

Now let us deduce from (10) that $f(e^{i\theta}) \in \Lambda_\alpha^p$. Let $0 < \rho < r < 1$ and $0 < \varphi - \theta < 2\pi$. Then

$$f(re^{i\varphi}) - f(re^{i\theta}) = \int_\Gamma f'(\zeta)\, d\zeta,$$

where the contour Γ goes radially from $re^{i\theta}$ to $\rho e^{i\theta}$, along $|z| = \rho$ to $\rho e^{i\varphi}$, then radially to $re^{i\varphi}$. It follows that

$$|f(re^{i\varphi}) - f(re^{i\theta})| \le \int_\rho^r |f'(se^{i\theta})|\, ds + \int_\rho^r |f'(se^{i\varphi})|\, ds$$

$$+ \int_\theta^\varphi |f'(\rho e^{it})|\, dt.$$

Now choose h, $0 < h < \frac{1}{2}$, and let $\varphi = \theta + h$. Applying Minkowski's inequality in both discrete and continuous form, we obtain

$$\left\{ \int_0^{2\pi} |f(re^{i(\theta+h)}) - f(re^{i\theta})|^p \, d\theta \right\}^{1/p} \leq 2 \int_\rho^r M_p(s, f') \, ds + h M_p(\rho, f'). \qquad (11)$$

The parameters r and ρ are still free. Assuming $r > 1 - h$, set $\rho = r - h$. Using the hypothesis (10), we conclude that for all $r > 1 - h$, the left-hand side of (11) is no larger than Ch^α, where the constant C is independent of r. Letting $r \to 1$, it follows that $f(e^{i\theta}) \in \Lambda_\alpha^p$.

To prove the converse, we begin with the estimate (3) for $|f'(z)|$. If $f(e^{i\theta}) \in \Lambda_\alpha^p$, Minkowski's inequality gives

$$M_p(r, f') \leq C \int_{-\pi}^\pi \frac{|t|^\alpha \, dt}{1 - 2r \cos t + r^2}.$$

Hence, as in the proof of Theorem 5.1, we obtain (10) for $\alpha < 1$.

If $\alpha = 1$, it must be shown that $f \in H^p$ and $f(e^{i\theta}) \in \Lambda_1^p$ imply $f' \in H^p$, $1 \leq p < \infty$. Consider first the case $p = 1$. Then $f \in H^1$ and by Lemma 1, $f(e^{i\theta})$ is equivalent to a function of bounded variation. Invoking Theorems 3.10 and 3.11, we conclude that $f(e^{i\theta})$ is absolutely continuous, $f' \in H^1$, and the radial limit $f'(e^{i\theta})$ is obtained by differentiation of $f(e^{i\theta})$. If it is also known that $f(e^{i\theta}) \in \Lambda_1^p$ for some $p > 1$, then

$$\int_0^{2\pi} \left| \frac{f(e^{i(\theta+h)}) - f(e^{i\theta})}{h} \right|^p \, d\theta \leq C.$$

It then follows by use of Fatou's lemma that the H^1 function $f'(z)$ has an L^p boundary function. Therefore $f' \in H^p$, by Theorem 2.11.

5.3. GROWTH OF A FUNCTION AND ITS DERIVATIVE

Our next object is to explore the relation between the mean growth of an analytic function and that of its derivative. We begin with a lemma which is especially useful in dealing with the case $p < 1$.

LEMMA 2. Any function $f \in H^p$ ($0 < p \leq \infty$) can be expressed in the form $f(z) = f_1(z) + f_2(z)$, where f_1 and f_2 are *nonvanishing* H^p functions such that $\|f_n\|_p \leq 2\|f\|_p$, $n = 1, 2$.

PROOF. If $f(z) \neq 0$, we may take $f_1 = f_2 = \frac{1}{2}f$. If f has zeros, we apply Theorem 2.5 to write $f(z) = B(z)g(z)$, where B is a Blaschke product and g is an H^p function with no zeros. Thus

$$f(z) = [B(z) - 1]g(z) + g(z).$$

THEOREM 5.5. Suppose $0 < p \leq \infty$ and $\beta > 0$, and let $f(z)$ be analytic in $|z| < 1$. Then

$$M_p(r, f) = O\left(\frac{1}{(1 - r)^\beta}\right) \tag{12}$$

if and only if

$$M_p(r, f') = O\left(\frac{1}{(1 - r)^{\beta+1}}\right). \tag{13}$$

PROOF. First let $1 \leq p \leq \infty$, and assume that (12) holds. By the Cauchy formula,

$$f'(z) = \frac{1}{2\pi i} \int_{|\zeta|=\rho} \frac{f(\zeta)\,d\zeta}{(\zeta - z)^2} = \frac{\rho}{2\pi} \int_0^{2\pi} \frac{f(\rho e^{i(t+\theta)})e^{i(t-\theta)}}{(\rho e^{it} - r)^2}\,dt,$$

where $\rho = \frac{1}{2}(1 + r)$. Minkowski's inequality (in continuous form) then gives

$$M_p(r, f') \leq \frac{1}{2\pi} \int_0^{2\pi} \frac{M_p(\rho, f)\,dt}{\rho^2 - 2\rho r \cos t + r^2}$$

$$= \frac{M_p(\rho, f)}{\rho^2 - r^2} = O\left(\frac{1}{(1 - r)^{\beta+1}}\right).$$

Conversely, if (13) holds, we apply Minkowski's inequality to the relation

$$|f(re^{i\theta})| \leq |f(0)| + \int_0^r |f'(se^{i\theta})|\,ds$$

and obtain

$$M_p(r, f) \leq |f(0)| + \int_0^r M_p(s, f')\,ds = O\left(\frac{1}{(1 - r)^\beta}\right).$$

We remark that the proof could also have been based on Jensen's inequality. Suppose now that $0 < p < 1$ and that

$$M_p(r, f) \leq \frac{C}{(1 - r)^\beta}.$$

Under the preliminary assumption that $f(z) \neq 0$ in $|z| < 1$, the function $F(z) = [f(z)]^p$ is analytic and

$$M_1(r, F) \leq \frac{C^p}{(1 - r)^{\beta p}}.$$

With $\rho = \frac{1}{2}(1 + r)$, this implies

$$M_1(r, F') \le \frac{M_1(\rho, F)}{\rho^2 - r^2} \le \frac{4C^p}{(1 - r)^{\beta p + 1}}.$$

On the other hand,

$$f'(z) = \frac{1}{p}[F(z)]^{1/p - 1}F'(z);$$

thus Hölder's inequality gives

$$M_p(r, f') \le \frac{1}{p}\{M_1(r, F)\}^{1/p - 1}M_1(r, F') \le \frac{(4/p)C}{(1 - r)^{\beta + 1}}.$$

If $f(z)$ has zeros, we fix ρ $(0 < \rho < 1)$ and appeal to the lemma to write

$$f(\rho z) = f_1(z) + f_2(z),$$

where f_1 and f_2 are nonvanishing H^p functions such that

$$\|f_n\|_p \le \frac{2C}{(1 - \rho)^\beta}, \qquad n = 1, 2,$$

and the constant C is independent of ρ. Hence, by the result just obtained,

$$M_p(r, f_n') \le \frac{(8/p)C}{(1 - \rho)^\beta(1 - r)}, \qquad n = 1, 2.$$

Since

$$\rho^p|f'(\rho z)|^p \le |f_1'(z)|^p + |f_2'(z)|^p,$$

it follows that

$$M_p(\rho^2, f') \le \frac{K}{(1 - \rho)^{\beta + 1}},$$

where K is independent of ρ. This easily gives (13).

The proof that (13) implies (12) when $0 < p < 1$ is much more difficult, and is omitted here (see *Notes*).

The theorem says, roughly, that f' grows faster than f by a factor $(1 - r)^{-1}$. The same phenomenon is reflected in the following more delicate result, which will be needed later in this chapter.

THEOREM 5.6. Let $f(z)$ be analytic in $|z| < 1$. Then for $1 < p < \infty$, $1 < a < \infty$, and $-1 < b < \infty$,

$$\int_0^1 (1 - r)^b\{M_p(r, f)\}^a \, dr \le C\left\{\int_0^1 (1 - r)^{a + b}\{M_p(r, f')\}^a \, dr + |f(0)|^a\right\},$$

where C is a constant independent of f.

PROOF. First suppose f is analytic in the closed unit disk, and integrate by parts:

$$\int_0^1 (1 - r)^b \{M_p(r, f)\}^a \, dr$$

$$= \frac{1}{b + 1} |f(0)|^a + \frac{1}{b + 1} \int_0^1 (1 - r)^{b+1} \frac{\partial}{\partial r} \{M_p(r, f)\}^a \, dr. \qquad (14)$$

But

$$\frac{\partial}{\partial r} \{M_p(r, f)\}^a = \frac{a}{p} \{M_p(r, f)\}^{a-p} \frac{\partial}{\partial r} \{M_p(r, f)\}^p,$$

and a calculation based on the Cauchy–Riemann equations gives

$$\left| \frac{\partial}{\partial r} |f(re^{i\theta})|^p \right| \leq p |f(re^{i\theta})|^{p-1} |f'(re^{i\theta})|.$$

Thus it follows from Hölder's inequality that

$$\left| \frac{\partial}{\partial r} \{M_p(r, f)\}^a \right| \leq a \{M_p(r, f)\}^{a-1} M_p(r, f').$$

Consequently, the integral on the right-hand side of (14) is dominated by

$$a \int_0^1 (1 - r)^{b+1} \{M_p(r, f)\}^{a-1} M_p(r, f') \, dr$$

$$\leq a \left\{ \int_0^1 (1 - r)^b \{M_p(r, f)\}^a \, dr \right\}^{1 - 1/a}$$

$$\times \left\{ \int_0^1 (1 - r)^{a+b} \{M_p(r, f')\}^a \, dr \right\}^{1/a},$$

where Hölder's inequality has been used again. This easily gives the desired inequality.

If $f(z)$ is not analytic in $|z| \leq 1$, replace it by $f(\rho z)$, where $0 < \rho < 1$, and let ρ tend to 1 to obtain the result by means of the Lebesgue monotone convergence theorem.

5.4. MORE ON CONJUGATE FUNCTIONS

We saw in Chapter 4 that the class h^p is preserved under harmonic conjugation if $1 < p < \infty$, but not if $0 < p \leq 1$ or if $p = \infty$. Nevertheless, the symmetry is restored in these latter cases if instead of the boundedness of the means one considers their order of growth. We shall confine attention to $1 \leq p \leq \infty$. The theorem remains true for $0 < p < 1$, but the proof is much more difficult.

THEOREM 5.7. Let $f(z) = u(z) + iv(z)$ be analytic in $|z| < 1$, and suppose

$$M_p(r, u) = O\left(\frac{1}{(1-r)^\beta}\right), \qquad 1 \le p \le \infty, \quad \beta > 0.$$

Then

$$M_p(r, v) = O\left(\frac{1}{(1-r)^\beta}\right).$$

PROOF. For $1 < p < \infty$, we could apply the theorem of M. Riesz, but we shall give a proof which makes no appeal to this deeper result. Let $\rho = \frac{1}{2}(1+r)$, and express $f(z)$ by the Poisson formula:

$$f(z) = \frac{1}{2\pi} \int_0^{2\pi} \frac{\rho e^{it} + z}{\rho e^{it} - z} u(\rho e^{it}) \, dt + i\gamma.$$

Then

$$|f'(re^{i\theta})| \le \frac{1}{\pi} \int_0^{2\pi} \frac{|u(\rho e^{i(\theta+t)})| \, dt}{\rho^2 - 2\rho r \cos t + r^2}.$$

By Minkowski's inequality,

$$M_p(r, f') \le \frac{1}{\pi} \int_0^{2\pi} \frac{M_p(\rho, u) \, dt}{\rho^2 - 2\rho r \cos t + r^2}$$

$$= \frac{2M_p(\rho, u)}{\rho^2 - r^2} = O\left(\frac{1}{(1-r)^{\beta+1}}\right).$$

Theorem 5.5 now gives the desired conclusion.

Generally speaking, then, a harmonic function and its conjugate have the same rate of growth. We are now in a position to show that they also have the same degree of smoothness on the boundary.

THEOREM 5.8. Let $f(z) = u(z) + iv(z)$ be analytic in $|z| < 1$, and suppose $u(z)$ is continuous in $|z| \le 1$. If $u(e^{i\theta}) \in \Lambda_\alpha$ ($\alpha < 1$), then $v(z)$ is continuous in $|z| \le 1$ and $v(e^{i\theta}) \in \Lambda_\alpha$. If $u(e^{i\theta}) \in \Lambda_*$, then $v(e^{i\theta}) \in \Lambda_*$.

PROOF. If we represent $f(z)$ as a Poisson integral of $u(e^{it})$ and follow the proof of Theorem 5.1, we see that the weaker condition $u(e^{i\theta}) \in \Lambda_\alpha$ still implies $f'(z) = O((1-r)^{\alpha-1})$. Thus $f(e^{i\theta}) \in \Lambda_\alpha$, by Theorem 5.1. To show that Λ_* is also preserved under conjugation, we may express $u(z)$ as the Poisson integral of $u(e^{it})$ and follow the proof of Theorem 5.3 to see that $u_{\theta\theta}(z) = O((1-r)^{-1})$.

Theorem 5.7 then implies $v_{\theta\theta}(z) = O((1 - r)^{-1})$, so that $f''(z) = O((1 - r)^{-1})$. Hence $f(e^{i\theta}) \in \Lambda_*$, by Theorem 5.3.

COROLLARY. Every function $\varphi(x)$ of class Λ_* has modulus of continuity $\omega(t) = O(t \log 1/t)$.

PROOF. Let $u(z)$ be the Poisson integral of the given function $\varphi \in \Lambda_*$, and let $f(z) = u(z) + iv(z)$ be its analytic completion. Then $f(z)$ is continuous in $|z| \leq 1$, and $f(e^{i\theta}) \in \Lambda_*$. The result therefore follows from the special case already noted as a corollary to Theorem 5.3.

5.5. COMPARATIVE GROWTH OF MEANS

If $f(z) \in H^p$ $(0 < p < \infty)$, it is possible to give a sharp estimate on the growth of $M_q(r, f)$ for any $q > p$. This theorem has a number of interesting applications. The proof will make use of the following lemma, which can be established by essentially the same argument used to prove the lemma in Section 4.6.

LEMMA 3. If $a > 1$ and $\rho = \frac{1}{2}(1 + r)$, then

$$\int_0^{2\pi} |\rho e^{it} - r|^{-a} \, dt = O((1 - r)^{1-a}), \qquad r \to 1.$$

THEOREM 5.9 (Hardy-Littlewood). Let $f(z)$ be analytic in $|z| < 1$, and suppose

$$M_p(r, f) \leq \frac{C}{(1 - r)^\beta}, \qquad 0 < p < \infty, \quad \beta \geq 0. \tag{15}$$

Then there is a constant $K = K(p, \beta)$ independent of f such that

$$M_q(r, f) \leq \frac{KC}{(1 - r)^{\beta + 1/p - 1/q}}, \qquad p < q \leq \infty. \tag{16}$$

If $\beta = 0$ (i.e., if $f \in H^p$), then $M_q(r, f) = o((1 - r)^{1/q - 1/p})$. The exponent $(\beta + 1/p - 1/q)$ is best possible in all cases.

PROOF. Let us first observe that it suffices to consider the case $q = \infty$. Indeed, suppose that (16) has been proved for $q = \infty$, and suppose for convenience that $K \geq 1$. Then for $p < q < \infty$,

$$M_q(r, f) = \left\{ \frac{1}{2\pi} \int_0^{2\pi} |f(re^{i\theta})|^p |f(re^{i\theta})|^{q-p} \, d\theta \right\}^{1/q}$$

$$\leq \{M_\infty(r, f)\}^{1 - p/q} \{M_p(r, f)\}^{p/q} \leq \frac{KC}{(1 - r)^\lambda},$$

where

$$\lambda = \beta p/q + (\beta + 1/p)(1 - p/q) = \beta + 1/p - 1/q.$$

Similar remarks apply to the "o" part of the theorem. Hence we may confine attention to the case $q = \infty$.

Suppose now that $1 < p < \infty$, and let p' be the conjugate index: $1/p + 1/p' = 1$. By the Cauchy formula,

$$f(z) = \frac{\rho}{2\pi} \int_0^{2\pi} \frac{f(\rho e^{it})e^{it}}{\rho e^{it} - z} \, dt, \qquad r < \rho < 1.$$

Applying Hölder's inequality and the hypothesis (15), we find

$$|f(re^{i\theta})| \le \frac{C}{(1 - \rho)^\beta} \left\{ \frac{1}{2\pi} \int_0^{2\pi} \frac{dt}{|\rho e^{it} - re^{i\theta}|^{p'}} \right\}^{1/p'}.$$

We now set $\rho = \frac{1}{2}(1 + r)$ and use Lemma 3 to obtain (16) for $q = \infty$. The proof for $p = 1$ is similar but easier.

Turning now to the "o" assertion, let $f \in H^p$, $1 \le p < \infty$. Given $\varepsilon > 0$, there exists $\delta > 0$ such that

$$\int_{-\delta}^{\delta} |f(e^{i(\theta + t)})|^p \, dt < \varepsilon^p \quad \text{for all } \theta.$$

On the other hand, the Cauchy formula gives

$$f(re^{i\theta}) = \frac{1}{2\pi} \int_{-\pi}^{\pi} \frac{f(e^{i(\theta + t)})e^{it}}{e^{it} - r} \, dt = \frac{1}{2\pi} \left\{ \int_{-\delta}^{\delta} + \int_{\delta}^{\pi} + \int_{-\pi}^{\delta} \right\}.$$

For each fixed δ, the last two integrals are obviously bounded as $r \to 1$. If $p > 1$, the integral over $(-\delta, \delta)$ is dominated by

$$\left\{ \int_{-\delta}^{\delta} |f(\theta + t)|^p \, dt \right\}^{1/p} \left\{ \int_{-\pi}^{\pi} \frac{dt}{|e^{it} - r|^{p'}} \right\}^{1/p'} < \frac{A\varepsilon}{(1 - r)^{1/p}},$$

where the lemma of Section 4.6 has been used. Similarly for $p = 1$. Hence $f(z) = o((1 - r)^{-1/p})$, which is what we wanted to show.

It remains to deal with the case $p < 1$. If $f(z) \ne 0$ in $|z| < 1$, the function $F(z) = [f(z)]^p$ is analytic, and (15) gives

$$M_1(r, F) = \{M_p(r, f)\}^p \le \frac{C^p}{(1 - r)^{\beta p}}.$$

Thus, by what we have already proved,

$$|f(z)|^p = |F(z)| \le \frac{KC^p}{(1 - r)^{\beta p + 1}},$$

which is the desired result. The same argument shows $f(z) = o((1 - r)^{-1/p})$ for nonvanishing H^p functions. If $f(z)$ has zeros, we fix $\rho < 1$ and use Lemma 2 to write

$$f(\rho z) = f_1(z) + f_2(z),$$

where f_1 and f_2 do not vanish and

$$\|f_n\|_p \leq 2M_p(\rho, f) \leq \frac{2C}{(1 - \rho)^\beta}, \qquad n = 1, 2.$$

Since $f_n(z) \neq 0$, it follows that

$$|f(\rho^2 e^{i\theta})| \leq |f_1(\rho e^{i\theta})| + |f_2(\rho e^{i\theta})| \leq \frac{4KC}{(1 - \rho)^{\beta + 1/p}}.$$

The constant K is independent of ρ. Finally, if $f \in H^p$, the general result $f(z) = o((1 - r)^{-1/p})$ follows directly from Lemma 2.

The function $f(z) = (1 - z)^{-\gamma}$, for suitable $\gamma > 0$, shows that the exponent $(\beta + 1/p - 1/q)$ cannot be improved. We leave the details to the reader.

The " o " assertion can also be reduced to the special case $p = 2$, which is easily proved as follows. Let $f(z) = \sum a_n z^n$ be in H^2, and fix N so that

$$\sum_{n=N}^{\infty} |a_n|^2 < \varepsilon^2.$$

Then

$$|f(z)| \leq \left| \sum_{n=0}^{N} a_n z^n \right| + \left\{ \sum_{n=N+1}^{\infty} |a_n|^2 \right\}^{1/2} \left\{ \sum_{n=N+1}^{\infty} r^{2n} \right\}^{1/2}$$

$$\leq O(1) + \varepsilon(1 - r)^{-1/2},$$

which shows that $f(z) = o((1 - r)^{-1/2})$.

It is natural to ask whether Theorem 5.9 has a converse. In particular, if $p < q$ and $M_q(r, f) \to \infty$ sufficiently slowly, must f be in H^p? Unfortunately, no such thing can be true, as the following theorem shows.

THEOREM 5.10. Let $\psi(r)$ be continuous and increasing on $0 \leq r < 1$, with $\psi(0) = 1$ and $\psi(r) \to \infty$ as $r \to 1$. Then there is a function $f(z)$ analytic in $|z| < 1$, satisfying $|f(z)| \leq \psi(|z|)$, yet having a radial limit on no set of positive measure. In particular, $f(z)$ is not of class H^p for any $p > 0$.

PROOF. The proof rests upon the following elementary observation.

LEMMA 4. Given a function $\psi(r)$ as in Theorem 5.10, there is a sequence of integers $0 < n_1 < n_2 < \cdots$ such that

$$\sum_{k=1}^{\infty} r^{n_k} \leq \psi(r), \qquad 0 \leq r < 1.$$

Momentarily taking the lemma for granted, let

$$f(z) = \sum_{k=1}^{\infty} \varepsilon_k z^{n_k}, \qquad \varepsilon_k = \pm 1.$$

Then $|f(z)| \leq \psi(r)$, and it follows from Theorem A.5 (in Appendix A) that for almost every choice of signs $\{\varepsilon_n\}$, $f(z)$ has a radial limit almost nowhere. This proves the theorem.

PROOF OF LEMMA. Let $r_0 = 0$, and let $r_1, r_2, \ldots$ be determined by $\psi(r_k) = k + 1$. Then $0 < r_1 < r_2 < \cdots < 1$. Choose n_1 such that $r_1^{n_1} \leq \frac{1}{2}$. Generally, select $n_k > n_{k-1}$ such that $r_k^{n_k} \leq 2^{-k}$. Given r in $[0, 1)$, let s be the index for which $r_{s-1} \leq r < r_s$. Then

$$\sum_{k=1}^{\infty} r^{n_k} \leq \sum_{k=1}^{\infty} r_s^{n_k} = \sum_{k=1}^{s-1} r_s^{n_k} + \sum_{k=s}^{\infty} r_s^{n_k}$$

$$\leq (s-1) + \sum_{k=s}^{\infty} r_k^{n_k} \leq (s-1) + 2^{-s+1}$$

$$\leq s = \psi(r_{s-1}) \leq \psi(r).$$

Although Theorem 5.9 is in a sense best possible, it can be slightly sharpened in one direction to produce a very useful result.

THEOREM 5.11 (Hardy–Littlewood). If $0 < p < q \leq \infty$, $f \in H^p$, $\lambda \geq p$, and $\alpha = 1/p - 1/q$, then

$$\int_0^1 (1 - r)^{\lambda\alpha - 1}\{M_q(r, f)\}^{\lambda} \, dr < \infty.$$

PROOF. It suffices to take $\lambda = p$, since by Theorem 5.9,

$$\int_0^1 (1 - r)^{\lambda\alpha - 1}\{M_q(r, f)\}^{\lambda} \, dr \leq C^{\lambda - p} \int_0^1 (1 - r)^{p\alpha - 1}\{M_q(r, f)\}^p \, dr.$$

Next, suppose the theorem were proved for $\lambda = p = 2$. Then if $f \in H^p$ $(0 < p < \infty)$ and $f(z) \neq 0$ in $|z| < 1$,

$$g(z) = [f(z)]^{p/2} \in H^2,$$

and the desired result (with $\lambda = p$) would follow. The general theorem could then be deduced by writing $f \in H^p$ as the sum of two non-vanishing H^p functions, as in Lemma 2.

Hence the theorem reduces to the special case $\lambda = p = 2$. To prove this, suppose first that $2 < q < \infty$, and let

$$f(z) = \sum_{n=0}^{\infty} a_n z^n \in H^2.$$

Then by Theorem 5.6,

$$\int_0^1 (1 - r)^{-2/q} \{M_q(r, f)\}^2 \, dr \leq C \left\{ |a_0|^2 + \int_0^1 (1 - r)^{2-2/q} \{M_q(r, f')\}^2 dr \right\}.$$

(17)

But it follows from Theorem 5.9 that

$$M_q(r, f') \leq K(1 - r)^{1/q - 1/2} M_2(r^{1/2}, f'),$$

so the second integral in (17) is dominated by a constant multiple of

$$\int_0^1 (1 - r)\{M_2(r^{1/2}, f')\}^2 \, dr = \sum_{n=1}^{\infty} n^2 |a_n|^2 \int_0^1 (1 - r) r^{n-1} \, dr$$

$$= \sum_{n=1}^{\infty} \frac{n}{n + 1} |a_n|^2 < \infty.$$

Finally, the result for $q = \infty$ is easily deduced from the case $q < \infty$ by Theorem 5.9.

5.6. FUNCTIONS WITH H^p DERIVATIVE

According to Theorem 3.11, every function which has an H^1 derivative is continuous in the closed disk and absolutely continuous on the boundary. In particular, $f' \in H^1$ implies $f \in H^\infty$. This latter result has an interesting generalization.

THEOREM 5.12 (Hardy–Littlewood). If $f' \in H^p$ for some $p < 1$, then $f \in H^q$, where $q = p/(1 - p)$. For each value of p, the index q is best possible.

PROOF. For convenience, assume $f(0) = 0$.

Case I: $\frac{1}{2} \leq p \leq 1$. Since $q \geq 1$, Minkowski's inequality may be applied (as in the proof of Theorem 5.5) to give

$$M_q(r, f) \leq \int_0^r M_q(s, f') \, ds.$$

But if $f' \in H^p$ and $q = p/(1 - p)$, it follows from Theorem 5.11 that

$$\int_0^1 M_q(s, f') \, ds < \infty.$$

Case II: $p < \frac{1}{2}$. Fix α in the range

$$1/2p < \alpha < 1/p;$$

then $\alpha > 1$. In view of Lemma 2, we may assume that $f'(z)$ has no zeros in $|z| < 1$. Letting $g(z) = [f'(z)]^{1/\alpha}$, we have

$$|f(re^{i\theta})| \leq \int_0^r |g(se^{i\theta})|^\alpha \, ds \leq [G(\theta)]^{\alpha-1} h(r, \theta),$$

where

$$G(\theta) = \sup_{0 < r < 1} |g(re^{i\theta})|$$

and

$$h(r, \theta) = \int_0^r |g(se^{i\theta})| \, ds.$$

Hence

$$|f(re^{i\theta})|^q \leq [G(\theta)]^{(\alpha-1)q} [h(r, \theta)]^q.$$

Now set

$$\beta = \frac{\alpha(1 - p)}{\alpha - 1} > 1,$$

and let $\beta' = \beta/(\beta - 1)$ be the conjugate index. By Hölder's inequality,

$$\{M_q(r, f)\}^q \leq \|G\|_{\alpha p}^{\alpha p/\beta} \left\{ \frac{1}{2\pi} \int_0^{2\pi} [h(r, \theta)]^{q\beta'} \, d\theta \right\}^{1/\beta'},$$

since $(\alpha - 1)q\beta = \alpha p$. By the Hardy–Littlewood maximal theorem (Theorem 1.9), $G \in L^{\alpha p}$. On the other hand,

$$q\beta' = \frac{\alpha p}{1 - \alpha p} > 1;$$

thus Minkowski's inequality gives

$$\left\{ \frac{1}{2\pi} \int_0^{2\pi} [h(r, \theta)]^{q\beta'} \, d\theta \right\}^{1/q\beta'} \leq \int_0^r M_{q\beta'}(s, g) \, ds,$$

which remains bounded as $r \to 1$, by Theorem 5.11. Thus $f \in H^q$.

To show that $q = p/(1 - p)$ cannot be replaced by any larger index, one need only consider $f'(z) = (1 - z)^{\varepsilon - 1/p}$ for small $\varepsilon > 0$.

EXERCISES

1. Let $f(z) = \sum_{n=0}^{\infty} a_n z^n$ be analytic in $|z| < 1$. Show that if

$$\sum_{n=1}^{N} n|a_n| = O(N^{1-\alpha}), \qquad 0 < \alpha \le 1,$$

then $f(z)$ is continuous in $|z| \le 1$ and $f(e^{i\theta}) \in \Lambda_\alpha$. Show further that the converse is true if $\alpha < 1$ and $a_n \ge 0$, $n = 1, 2, \ldots$.

2. Let $f(z) = \sum_{n=0}^{\infty} a_n z^n$ be analytic in $|z| < 1$. Show that if

$$\sum_{n=1}^{N} n|a_n| = O(\log N),$$

then $f(z)$ is continuous in $|z| \le 1$ and $f(e^{i\theta})$ has modulus of continuity $\omega(t) = O(t \log 1/t)$.

3. Show that if

$$\sum_{n=1}^{N} n^2 |a_n| = O(N),$$

then $f(z)$ is continuous in $|z| \le 1$ and $f(e^{i\theta}) \in \Lambda_*$. Show that the converse is true if $a_n \ge 0$, $n = 1, 2, \ldots$.

4. Show that if $f(z)$ is continuous in $|z| \le 1$ and $f(e^{i\theta}) \in \Lambda_\alpha$, $0 < \alpha \le 1$, then

$$\sum_{n=1}^{N} n^2 |a_n|^2 = O(N^{2(1-\alpha)}).$$

If $f(e^{i\theta}) \in \Lambda_*$, show that

$$\sum_{n=1}^{N} n^3 |a_n|^2 = O(N);$$

hence

$$\sum_{n=1}^{N} n^2 |a_n| = O(N^{3/2}).$$

(Duren, Shapiro, and Shields [1]).

5. Show that if m is an integer greater than 1, and if $0 < \alpha < 1$, the function

$$f(z) = \sum_{k=1}^{\infty} m^{-k\alpha} z^{m^k}$$

is continuous in $|z| \le 1$ and $f(e^{i\theta}) \in \Lambda_\alpha$.

6. Show that the first two estimates in Exercise 4 are best possible in the sense that "O" cannot be replaced by "o." (The third estimate is also best possible; see Duren, Shapiro, and Shields [1].)

7. Show that if $f(z)$ is analytic in $|z| < 1$, continuous in $|z| \leq 1$, and $f(e^{i\theta}) \in \Lambda_\alpha$ for some $\alpha > \frac{1}{2}$, then $\sum_{n=0}^\infty |a_n| < \infty$. Show further that the index $\frac{1}{2}$ is best possible: the series $\sum_{n=1}^\infty |a_n|$ may diverge if only $f(e^{i\theta}) \in \Lambda_{1/2}$ (Theorem of S. Bernstein).

8. Show that if $f(z)$ is continuous in $|z| \leq 1$ and $f(e^{i\theta}) \in \Lambda_\alpha$, $0 < \alpha \leq 1$, then $\sum_{n=1}^\infty |a_n|^\gamma < \infty$ for all $\gamma > 2/(2\alpha + 1)$. (*Hint*: Show that $\sum n^\beta |a_n|^2 < \infty$ for all $\beta < 2\alpha$.)

9. Show that if $f' \in H^p$, $1 < p \leq \infty$, then $f(z)$ is continuous in $|z| \leq 1$ and $f(e^{i\theta}) \in \Lambda_\alpha$, $\alpha = 1 - 1/p$.

10. Show that if $\varphi(t)$ is continuous and

$$|\varphi(t + h) - 2\varphi(t) + \varphi(t - h)| \leq Ah^\alpha, \qquad 0 < \alpha < 1,$$

then $\varphi \in \Lambda_\alpha$. (*Suggestion*: Imitate the proof of Theorem 5.1.)

11. For $1 \leq p < \infty$, show that the class Λ_α^p contains only the constant functions if $\alpha > 1$.

12. Let $f(z) = u(z) + iv(z)$ be analytic in $|z| < 1$. Suppose $u \in h^p$ $(1 < p < \infty)$ and $u(e^{i\theta}) \in \Lambda_\alpha^p$ $(0 < \alpha < 1)$. Show that $v(e^{i\theta}) \in \Lambda_\alpha^p$. In other words, show that the class Λ_α^p is self-conjugate.

13. Prove that if $f \in H^p$ $(1 < p < \infty)$ and $f(e^{i\theta}) \in \Lambda_\alpha^p$ $(1/p < \alpha \leq 1)$, then $f(z)$ is continuous in $|z| \leq 1$ and $f(e^{i\theta}) \in \Lambda_\beta$, $\beta = \alpha - 1/p$.

14. Show that the function

$$\varphi(x) = x \log 1/x, \qquad x > 0,$$

belongs to Λ_* but not to Λ_1.

15. Prove Theorem 5.2.

16. Using the example $f(z) = (1 - z)^{-\gamma}$, show that the exponent $(\beta + 1/p - 1/q)$ in Theorem 5.9 is best possible.

NOTES

Almost everything in this chapter is due to Hardy and Littlewood. Lemma 1 is in their paper [2]. Theorems 5.1, 5.4–5.6, 5.9, and 5.11 occur in their paper [5]. The proofs of Theorems 5.5 and 5.7 for the case $p < 1$ may be found in their paper [6]. The simple but enormously useful remark that every H^p function is the sum of two nonvanishing ones (Lemma 2) can be traced to their paper [1]. Further results and generalizations are in their papers [5] and [8]. The "o" growth condition given in Theorem 5.9 is best possible; see G. D. Taylor [1] and Duren and Taylor [1]. Flett [2] has recently based a proof of Theorem 5.11 on the Marcinkiewicz interpolation theorem. Zygmund [2, 4] introduced the class Λ_* and proved Theorem 5.3 in somewhat different form. The self-conjugacy of Λ_α (Theorem 5.8) goes back to Privalov [1]. Zygmund [2] showed that Λ_* is self-conjugate. For a direct proof that every Λ_* function has modulus of continuity which is $O(t \log 1/t)$, see Zygmund [4], Vol. I, p. 44.

The converse to Theorem 5.12 is totally false. The "Bloch–Nevanlinna conjecture" asserted that if $f \in N$, then $f' \in N$, but this has been disproved in many ways. In fact, there exist functions $f \in H^\infty$ continuous in the closed disk, such that $f'(z)$ has a radial limit almost nowhere. See, for example, Lohwater, Piranian, and Rudin [1] and Duren [4]. Hayman [3] has recently shown that $f' \in N$ does not imply $f \in N$, an unexpected result in view of Theorem 5.12. Caughran [2] has obtained results comparing the canonical factorizations of f and f' in case $f' \in H^1$.

If a function $f(z) = \sum a_n z^n$ belongs to a certain H^p space, what can be said about its Taylor coefficients a_n? Clearly, one can hope to describe only the "eventual behavior" of $\{a_n\}$ as $n \to \infty$, since any finite number of coefficients can be changed arbitrarily without upsetting the fact that f is in H^p. It is also interesting to ask how an H^p function can be recognized by the behavior of its Taylor coefficients. Ideally, one would like to find a condition on the a_n which is both necessary and sufficient for f to be in H^p. For $p = 2$, of course, the problem is completely solved: $f \in H^2$ if and only if $\sum |a_n|^2 < \infty$. But the general situation is much more complicated, and no complete answer is available. If $1 < p < \infty$, the problem is equivalent to that of describing the Fourier coefficients of L^p functions, as the M. Riesz theorem shows.

This chapter contains some scattered information about the coefficients of H^p functions. Curiously, the results are most complete in the case $0 < p < 1$.

6.1. HAUSDORFF–YOUNG INEQUALITIES

Any available information about the Fourier coefficients of L^p functions can be applied, in particular, to the Taylor coefficients of H^p functions. In fact, the two sets of coefficient sequences are essentially the same if $1 < p < \infty$. The Hausdorff–Young theorem states that if $\varphi(x) \in L^p = L^p[0, 2\pi]$, $1 < p \leq 2$, then its sequence $\{c_n\}$ of Fourier coefficients is in ℓ^q $(1/p + 1/q = 1)$ and

$$\left\{ \sum_{n=-\infty}^{\infty} |c_n|^q \right\}^{1/q} \leq \left\{ \frac{1}{2\pi} \int_0^{2\pi} |\varphi(x)|^p \, dx \right\}^{1/p}.$$

Conversely, every ℓ^p sequence $\{c_n\}$ of complex numbers $(1 < p \leq 2)$ is the sequence of Fourier coefficients of some $\varphi \in L^q$ $(1/p + 1/q = 1)$, and

$$\left\{ \frac{1}{2\pi} \int_0^{2\pi} |\varphi(x)|^q \, dx \right\}^{1/q} \leq \left\{ \sum_{n=-\infty}^{\infty} |c_n|^p \right\}^{1/p}.$$

This result may also be expressed in H^p language, as follows.

THEOREM 6.1. If

$$f(z) = \sum_{n=0}^{\infty} a_n z^n \in H^p \qquad (1 \leq p \leq 2),$$

then $\{a_n\} \in \ell^q$ $(1/p + 1/q = 1)$ and

$$\|\{a_n\}\|_q \leq \|f\|_p. \tag{1}$$

Conversely, if $\{a_n\}$ is any ℓ^p sequence of complex numbers $(1 \leq p \leq 2)$, then $f(z) = \sum a_n z^n$ is in H^q $(1/p + 1/q = 1)$ and

$$\|f\|_q \leq \|\{a_n\}\|_p. \tag{2}$$

PROOF. The first statement follows easily from the Hausdorff–Young theorem. Indeed, if $f \in H^p$, then the radial limit $f(e^{i\theta})$ is in L^p and $\{a_n\}$ is its sequence of Fourier coefficients. On the other hand, if $\{a_n\} \in \ell^p$ $(1 \leq p \leq 2)$, then $f \in H^2$, and the numbers a_n are the Fourier coefficients of $f(e^{i\theta})$. The Hausdorff–Young theorem then tells us that $f(e^{i\theta}) \in L^q$, which implies $f \in H^q$, and (2) follows.

Neither part of the theorem remains valid if $p > 2$. In fact, the hypothesis that $f \in H^p$ for some $p > 2$ implies nothing more about the $|a_n|$ than $\{a_n\} \in \ell^2$. And if the condition $\{a_n\} \in \ell^2$ is weakened to $\{a_n\} \in \ell^p$ for some $p > 2$, nothing reasonable can be said about $f(z)$. These claims are justified by Theorem A.5 of Appendix A. According to this theorem, the mere assumption that $\{a_n\} \in \ell^2$ implies that, for almost every choice of signs $\{\varepsilon_n\}$, the function $\sum \varepsilon_n a_n z^n$ is in H^p for all $p < \infty$. On the other hand, if $\{a_n\} \notin \ell^2$, almost every choice of signs produces a function $\sum \varepsilon_n a_n z^n$ having a radial limit almost nowhere.

6.2. THEOREM OF HARDY AND LITTLEWOOD

The following theorem provides further information about the coefficients of H^p functions.

THEOREM 6.2 (Hardy–Littlewood). If

$$f(z) = \sum_{n=0}^{\infty} a_n z^n \in H^p, \qquad 0 < p \le 2,$$

then $\sum n^{p-2} |a_n|^p < \infty$ and

$$\left\{ \sum_{n=0}^{\infty} (n+1)^{p-2} |a_n|^p \right\}^{1/p} \le C_p \|f\|_p, \tag{3}$$

where C_p depends only on p.

PROOF. We suppose $1 \le p \le 2$, postponing the case $0 < p < 1$ to Section 6.4. If $p = 1$, the theorem reduces to Hardy's inequality (Corollary to Theorem 3.15). If $p = 2$, it follows from the Parseval relation. A proof for intermediate values of p can be based on the Marcinkiewicz interpolation theorem, but the argument given here will be self-contained.

Let μ be the measure defined on the set of integers by

$$\mu(n) = (|n| + 1)^{-2}, \qquad n = 0, \pm 1, \pm 2, \dots.$$

If $g \in L^2$ and

$$g(t) \sim \sum_{n=-\infty}^{\infty} b_n e^{int},$$

let

$$\tilde{g}(n) = (|n| + 1) b_n, \qquad n = 0, \pm 1, \pm 2, \dots.$$

For $s > 0$, let

$$E_s = \{ n : |\tilde{g}(n)| > s \}.$$

Then

$$s^2 \mu(E_s) \le \sum_{n=-\infty}^{\infty} |\tilde{g}(n)|^2 \mu(n) = \sum_{n=-\infty}^{\infty} |b_n|^2 = \|g\|_2^2. \tag{4}$$

Furthermore, since $|b_n| \le \|g\|_1$,

$$E_s \subset F_s = \{ n : (|n| + 1) \|g\|_1 > s \};$$

thus

$$\mu(E_s) \le \mu(F_s) = \sum_{n \in F_s} (|n| + 1)^{-2} \le C s^{-1} \|g\|_1. \tag{5}$$

Now write

$$g(t) = \varphi_s(t) + \psi_s(t),$$

where

$$\varphi_s(t) = \begin{cases} g(t), & |g(t)| \geq s \\ 0, & |g(t)| < s. \end{cases}$$

Then

$$\sum_{n=-\infty}^{\infty} (|n| + 1)^{p-2}|b_n|^p = \sum_{n=-\infty}^{\infty} |\tilde{g}(n)|^p\mu(n)$$

$$\leq 2^p \left\{ \sum_{n=-\infty}^{\infty} |\tilde{\varphi}_s(n)|^p\mu(n) + \sum_{n=-\infty}^{\infty} |\tilde{\psi}_s(n)|^p\mu(n) \right\}. \tag{6}$$

Let

$$\alpha(s) = \mu(\{n : |\tilde{\varphi}_s(n)| > s\}),$$
$$\beta(s) = \mu(\{n : |\tilde{\psi}_s(n)| > s\}).$$

Then in view of (4) and (5),

$$\sum_{n=-\infty}^{\infty} |\tilde{\varphi}_s(n)|^p\mu(n) = -\int_0^\infty s^p d\alpha(s) = p \int_0^\infty s^{p-1}\alpha(s)\, ds$$

$$\leq pC \int_0^\infty s^{p-2}\|\varphi_s\|_1\, ds = \frac{pC}{2\pi} \int_0^\infty s^{p-2} \int_0^{2\pi} |\varphi_s(t)|\, dt\, ds$$

$$= \frac{pC}{2\pi} \int_0^{2\pi} |g(t)| \int_0^{|g(t)|} s^{p-2}\, ds\, dt = A_p \int_0^{2\pi} |g(t)|^p\, dt,$$

where A_p is a constant depending only on p. Similarly, by (4),

$$\sum_{n=-\infty}^{\infty} |\tilde{\psi}_s(n)|^p\mu(n) = -\int_0^\infty s^p\, d\beta(s) = p \int_0^\infty s^{p-1}\beta(s)\, ds$$

$$\leq p \int_0^\infty s^{p-3}\|\psi_s\|_2{}^2\, ds = \frac{p}{2\pi} \int_0^\infty s^{p-3} \int_0^{2\pi} |\psi_s(t)|^2\, dt\, ds$$

$$= \frac{p}{2\pi} \int_0^{2\pi} |g(t)|^2 \int_{|g(t)|}^\infty s^{p-3}\, ds\, dt \leq B_p \int_0^{2\pi} |g(t)|^p\, dt.$$

Combining these two estimates with (6), we have

$$\left\{ \sum_{n=-\infty}^{\infty} (|n| + 1)^{p-2}|b_n|^p \right\}^{1/p} \leq C_p\|g\|_p \tag{7}$$

if $g \in L^2$.

If $f \in H^p$ $(1 < p < 2)$, the desired estimate (3) now follows after approximating the boundary function $f(e^{it})$ by

$$g_M(t) = \begin{cases} f(e^{it}), & |f(e^{it})| \le M, \\ 0, & |f(e^{it})| > M, \end{cases}$$

applying (7) to g_M, and letting $M \to \infty$.

As with the Hausdorff–Young theorem, the converse to Theorem 6.2 is false if $p < 2$. However, the converse is true for indices larger than 2, and can be deduced from Theorem 6.2 by a duality argument. The exact statement is as follows.

THEOREM 6.3. Let $\{a_n\}$ be a sequence of complex numbers such that $\sum n^{q-2}|a_n|^q < \infty$ for some q, $2 \le q < \infty$. Then the function $f(z) = \sum_{n=0}^{\infty} a_n z^n$ is in H^q, and

$$\|f\|_q \le C_q \left\{ \sum_{n=0}^{\infty} (n+1)^{q-2}|a_n|^q \right\}^{1/q},$$

where C_q depends only on q.

PROOF. Let $p = q/(q-1)$, and let

$$G(e^{i\theta}) = \sum_{k=-n}^{n} c_k e^{ik\theta}$$

be an arbitrary trigonometric polynomial with $\|G\|_p \le 1$. By the M. Riesz theorem, its "analytic projection"

$$g(e^{i\theta}) = \sum_{k=0}^{n} c_k e^{ik\theta}$$

satisfies $\|g\|_p \le A_p \|G\|_p \le A_p$. Let

$$s_n(z) = \sum_{k=0}^{n} a_k z^k.$$

Then

$$\left| \frac{1}{2\pi} \int_0^{2\pi} G(e^{i\theta}) \overline{s_n(re^{i\theta})} \, d\theta \right| \le \sum_{k=0}^{n} |c_k| \, |a_k|$$

$$= \sum_{k=0}^{n} |c_k|(k+1)^{(p-2)/p}|a_k|(k+1)^{(q-2)/q}$$

$$\le \left\{ \sum_{k=0}^{n} (k+1)^{p-2}|c_k|^p \right\}^{1/p} \left\{ \sum_{k=0}^{n} (k+1)^{q-2}|a_k|^q \right\}^{1/q}$$

$$\le C_p \|g\|_p \left\{ \sum_{k=0}^{n} (k+1)^{q-2}|a_k|^q \right\}^{1/q} = C_p A_p \left\{ \sum_{k=0}^{\infty} (k+1)^{q-2}|a_k|^q \right\}^{1/q}.$$

Taking the supremum over all G with $\|G\|_p \le 1$, we have

$$M_q(r, s_n) \le C_p A_p \left\{ \sum_{k=0}^{\infty} (k+1)^{q-2} |a_k|^q \right\}^{1/q},$$

and the result follows by letting $n \to \infty$.

6.3. THE CASE $p \le 1$

As pointed out in the corollary to Theorem 3.4, the Taylor coefficients of an H^1 function must tend to zero, by the Riemann–Lebesgue lemma. The simple example $(1-z)^{-1}$ shows this is false for H^p functions with $p < 1$, but a sharp asymptotic estimate for the coefficients can be given as follows.

THEOREM 6.4. If

$$f(z) = \sum_{n=0}^{\infty} a_n z^n \in H^p, \qquad 0 < p \le 1,$$

then

$$a_n = o(n^{1/p-1}) \tag{8}$$

and

$$|a_n| \le C n^{1/p-1} \|f\|_p. \tag{9}$$

Furthermore, the estimate (8) is best possible for each p: given any positive sequence $\{\delta_n\}$ tending to zero, there exists $f \in H^p$ such that

$$a_n \ne O(\delta_n n^{1/p-1}).$$

PROOF. We have already observed that $a_n = o(1)$ if $f \in H^1$. This statement cannot be improved, even if $f \in H^\infty$. Indeed, given a sequence $\{\delta_n\}$ tending to zero, choose integers $0 < n_1 < n_2 < \cdots$ such that $\delta_{n_k} < 2^{-k}$. Define

$$a_n = \begin{cases} k\,\delta_{n_k} & \text{if } n = n_k, \\ 0 & \text{otherwise.} \end{cases}$$

Then $\sum |a_n| < \infty$, so $f(z) = \sum a_n z^n$ is continuous in $|z| \le 1$, yet $a_n \ne O(\delta_n)$. Now suppose $f \in H^p$, $p < 1$. The coefficient a_n has the representation

$$a_n = \frac{1}{2\pi i} \int_{|z|=r} \frac{f(z)}{z^{n+1}} \, dz, \qquad 0 < r < 1.$$

Hence

$$|a_n| \le r^{-n} M_1(r, f), \qquad 0 < r < 1.$$

But by Theorem 5.9,

$$M_1(r, f) = o((1 - r)^{1 - 1/p}).$$

With the choice $r = 1 - 1/n$, (8) follows. Theorem 5.9 also gives (9) in the same way.

Functions of the form $(1 - z)^{-\gamma}$ show the exponent in (8) cannot be reduced. The proof that the estimate is best possible will be postponed to Section 6.4.

6.4. MULTIPLIERS

A complex sequence $\{\lambda_n\}$ is said to be a *multiplier* of H^p into the sequence space ℓ^q if $\{\lambda_n a_n\} \in \ell^q$ whenever $\sum a_n z^n \in H^p$. Similarly, $\{\lambda_n\}$ is a multiplier of H^p into H^q if $\sum a_n z^n \in H^p$ implies $\sum \lambda_n a_n z^n \in H^q$. One way to give information about the coefficients of H^p functions is to identify the multipliers of H^p into various spaces. Results of this type have a number of interesting consequences. We shall content ourselves with describing the multipliers of H^p $(0 < p < 1)$ into ℓ^q $(p \le q \le \infty)$ and of H^1 into ℓ^1 and into ℓ^2 (alias H^2). Some preliminary remarks are in order.

First of all, even though H^p and ℓ^p are not Banach spaces if $p < 1$, they are complete topological vector spaces whose topologies are induced by the translation invariant metrics $\|f - g\|_p^p$ and

$$\|x - y\|_p^p = \sum |x_n - y_n|^p,$$

respectively (see Section 3.2). In other words, H^p and ℓ^p are *F-spaces*, in the terminology introduced by Banach [1]. The closed graph theorem therefore applies to mappings between these spaces (see Banach [1] or Dunford and Schwartz [1], Chap. II).

A sequence $\{\lambda_n\}$ may be regarded as inducing a linear operator

$$\Lambda : \sum a_n z^n \to \{\lambda_n a_n\}$$

from H^p to ℓ^q. The *domain* of Λ is the linear manifold

$$\mathscr{D}(\Lambda) = \{f \in H^p : \Lambda(f) \in \ell^q\}.$$

If $p < \infty$, $\mathscr{D}(\Lambda)$ is dense in H^p, since it contains all polynomials. Furthermore, it is clear that $\mathscr{D}(\Lambda) = H^p$ if and only if $\{\lambda_n\}$ is a multiplier of H^p into ℓ^q. The closed graph theorem therefore tells us that Λ is a bounded operator from H^p to ℓ^q if $\{\lambda_n\}$ is a multiplier. We have only to check that Λ is closed.

To show Λ is a closed operator, suppose

$$f_k \in \mathscr{D}(\Lambda), \qquad f_k \to f \in H^p,$$

and

$$g_k = \Lambda(f_k) \to g \in \ell^q.$$

Let

$$f_k(z) = \sum a_n^{(k)} z^n, \quad f(z) = \sum a_n z^n, \quad \text{and} \quad g = \{b_n\}.$$

Then by Theorem 6.4, $a_n^{(k)} \to a_n$, $n = 0, 1, \ldots$. Since also $\lambda_n a_n^{(k)} \to b_n$, this shows that $\lambda_n a_n = b_n$, $n = 0, 1, \ldots$. In other words, $f \in \mathscr{D}(\Lambda)$ and $\Lambda(f) = g$, which proves that Λ is closed.

THEOREM 6.5. Suppose $0 < p \le 1$. Then $\{\lambda_n\}$ is a multiplier of H^p into ℓ^∞ if and only if

$$\lambda_n = O(n^{1-1/p}). \tag{10}$$

PROOF. If $f(z) = \sum a_n z^n$ is in H^p and (10) holds, then $\{\lambda_n a_n\} \in \ell^\infty$, by Theorem 6.4. (In fact, $\lambda_n a_n \to 0$.) Conversely, if $\{\lambda_n\}$ is a multiplier, then by the closed graph theorem (as discussed above), Λ is a bounded operator from H^p to ℓ^∞. In other words,

$$\sup_n |\lambda_n a_n| \le C\|f\|_p, \quad f \in H^p. \tag{11}$$

Now let

$$g(z) = (1 - z)^{-1-1/p} = \sum b_n z^n,$$

where $b_n \sim Bn^{1/p}$; and choose $f(z) = g(rz)$ for fixed $r < 1$. Then by (11) and the lemma in Section 4.6,

$$|\lambda_n| n^{1/p} r^n \le C(1 - r)^{-1}.$$

The choice $r = 1 - 1/n$ now gives (10).

As a corollary, we can now show that the estimate (8) in Theorem 6.4 is best possible.

COROLLARY. If $\{d_n\}$ is a sequence of positive numbers such that $a_n = O(d_n)$ for every function $\sum a_n z^n$ in H^p $(p \le 1)$, then there is an $\varepsilon > 0$ such that

$$d_n n^{1-1/p} \ge \varepsilon, \quad n = 1, 2, \ldots .$$

PROOF. If $a_n = O(d_n)$ for every $f \in H^p$, then $\{1/d_n\}$ multiplies H^p into ℓ^∞. Thus $1/d_n = O(n^{1-1/p})$, as claimed.

THEOREM 6.6. Suppose $0 < p < 1$ and $p \le q < \infty$. Then $\{\lambda_n\}$ is a multiplier of H^p into ℓ^q if and only if

$$\sum_{n=1}^{N} n^{q/p} |\lambda_n|^q = O(N^q). \tag{12}$$

Before passing to the proof, let us apply the result to establish the Hardy–Littlewood theorem (Theorem 6.2) for $p < 1$.

COROLLARY. If $\sum a_n z^n \in H^p$ $(0 < p < 1)$, then

$$\sum_{n=1}^{\infty} n^{p-2} |a_n|^p < \infty.$$

PROOF. Let $q = p$, and observe that $\lambda_n = n^{1-2/p}$ then satisfies (12). Hence $\{n^{1-2/p}\}$ is a multiplier of H^p into ℓ^p. The closed graph theorem now gives the inequality (3) stated in Theorem 6.2.

The following lemma will be needed in the proof of Theorem 6.6.

LEMMA. Suppose $b_n \geq 0$ and $0 < \beta < \alpha$. Then

$$\sum_{n=1}^{N} n^\alpha b_n = O(N^\beta) \tag{13}$$

if and only if

$$\sum_{n=N}^{\infty} b_n = O(N^{\beta - \alpha}). \tag{14}$$

PROOF. Let (13) hold and let $S_n = \sum_{k=1}^{n} k^\alpha b_k$. Then

$$\sum_{n=N}^{M} b_n = \sum_{n=N}^{M-1} S_n [n^{-\alpha} - (n+1)^{-\alpha}] + S_M M^{-\alpha} - S_{N-1} N^{-\alpha}$$

$$\leq C \sum_{n=N}^{M-1} n^{\beta - \alpha - 1} + CM^{\beta - \alpha}.$$

Letting $M \to \infty$, we obtain (14). Conversely, assume (14) and let $R_n = \sum_{k=n}^{\infty} b_k$. Then

$$\sum_{n=1}^{N} n^\alpha b_n = \sum_{n=1}^{N} [n^\alpha - (n-1)^\alpha] R_n - N^\alpha R_{N+1} \leq CN^\beta.$$

PROOF OF THEOREM 6.6. By the lemma, the condition (12) implies

$$\sum_{n=N}^{\infty} |\lambda_n|^q = O(N^{q(1-1/p)}). \tag{15}$$

Assume without loss of generality that $\lambda_n \geq 0$ and $\sum_{n=1}^{\infty} \lambda_n^q = 1$. Let $s_1 = 0$ and

$$s_n = 1 - \left\{ \sum_{k=n}^{\infty} \lambda_k^q \right\}^{1/\gamma}, \qquad n = 2, 3, \ldots,$$

where $\gamma = q(1/p - 1)$. Note that s_n increases to 1 as $n \to \infty$. By Theorem 5.11, $f \in H^p$ $(0 < p < 1)$ implies

$$\int_0^1 (1 - r)^{\gamma - 1} M_1{}^q(r, f) \, dr < \infty, \qquad p \le q < \infty.$$

Since $M_1(r, f) \ge r^n|a_n|$, where $f(z) = \sum a_n z^n$, it follows that

$$\infty > \sum_{n=1}^{\infty} \int_{s_n}^{s_{n+1}} (1 - r)^{\gamma - 1} M_1{}^q(r, f) \, dr$$

$$\ge \sum_{n=1}^{\infty} |a_n|^q \int_{s_n}^{s_{n+1}} (1 - r)^{\gamma - 1} r^{nq} \, dr$$

$$\ge \frac{1}{\gamma} \sum_{n=1}^{\infty} |a_n|^q (s_n)^{nq} \{(1 - s_n)^\gamma - (1 - s_{n+1})^\gamma\}$$

$$= \frac{1}{\gamma} \sum_{n=1}^{\infty} \lambda_n{}^q |a_n|^q (s_n)^{nq},$$

by the definition of s_n. But by (15),

$$\left\{\sum_{k=n}^{\infty} \lambda_n{}^q\right\}^{1/\gamma} \le \frac{C}{n},$$

which shows that

$$(s_n)^{nq} \ge (1 - C/n)^{nq} \to e^{-Cq} > 0.$$

Since these factors $(s_n)^{nq}$ are eventually bounded away from zero, we have shown that $\{\lambda_n\}$ is a multiplier of H^p into ℓ^q if it satisfies (12).

Conversely, if $\{\lambda_n\}$ is a multiplier, then by the closed graph theorem,

$$\left\{\sum_{n=0}^{\infty} |\lambda_n a_n|^q\right\}^{1/q} \le C\|f\|_p.$$

Choosing $f(z) = g(rz)$ as in the proof of Theorem 6.5, we now find

$$\sum_{n=1}^{\infty} n^{q/p} |\lambda_n|^q r^{nq} \le C(1 - r)^{-q}.$$

Hence

$$r^{Nq} \sum_{n=1}^{N} n^{q/p} |\lambda_n|^q \le C(1 - r)^{-q},$$

and (12) follows upon setting $r = 1 - 1/N$. Note that the argument shows (12) is necessary even if $p \ge 1$ or $q < p$.

The condition (12) also characterizes the multipliers of H^1 into ℓ^q, if $2 \le q < \infty$. We give the proof first in the case $q = 2$.

THEOREM 6.7. The sequence $\{\lambda_n\}$ is a multiplier of H^1 into H^2 if and only if

$$\sum_{n=1}^{N} n^2|\lambda_n|^2 = O(N^2). \tag{16}$$

PROOF. The necessity of (16) was observed in the proof of Theorem 6.6. In proving the sufficiency, it is enough to consider functions without zeros, since by the lemma in Section 5.3, every H^1 function is the sum of two non-vanishing H^1 functions. Thus suppose $f(z) = \sum a_n z^n$ is an H^1 function which does not vanish in $|z| < 1$, and write $f = \varphi^2$, where $\varphi(z) = \sum b_n z^n$ is in H^2. Then

$$a_n = \sum_{k=0}^{n} b_k b_{n-k}.$$

It is to be shown that (16) and $\sum |b_n|^2 < \infty$ imply $\sum |\lambda_n a_n|^2 < \infty$. Without loss of generality, suppose $\lambda_n \geq 0$ and $b_n \geq 0$. Then

$$a_n \leq 2 \sum_{k=[n/2]}^{n} b_k b_{n-k},$$

where $[x]$ denotes the greatest integer $\leq x$. Now apply the Cauchy–Schwarz inequality:

$$a_n^2 \leq 4 \sum_{k=0}^{\infty} b_k^2 \sum_{k=[n/2]}^{n} b_k^2 = C \sum_{k=[n/2]}^{n} b_k^2 = C\beta_n,$$

say. It follows that

$$\sum_{n=1}^{N} \lambda_n^2 a_n^2 \leq C \sum_{n=1}^{N} \lambda_n^2 \beta_n.$$

But a summation by parts gives

$$\sum_{n=1}^{N} \lambda_n^2 \beta_n = N^{-2} S_N \beta_N - \sum_{n=1}^{N-1} S_n \Delta(n^{-2}\beta_n),$$

where

$$S_N = \sum_{n=1}^{N} n^2 \lambda_n^2$$

and $\Delta x_n = x_{n+1} - x_n$. The term $N^{-2}S_N \beta_N$ is bounded, by (16) and the assumption $\sum b_k^2 < \infty$, so it remains only to show that

$$\sum_{n=1}^{\infty} n^2 |\Delta(n^{-2}\beta_n)| < \infty. \tag{17}$$

But

$$\Delta(n^{-2}\beta_n) = (n+1)^{-2} \Delta\beta_n + \beta_n \Delta(n^{-2}).$$

Since $\sum b_k{}^2 < \infty$, it is clear that

$$\sum_{n=1}^{\infty} \left(\frac{n}{n+1}\right)^2 |\Delta \beta_n| < \infty;$$

and

$$\sum_{n=1}^{\infty} n^2 \beta_n |\Delta(n^{-2})| \le 2 \sum_{n=1}^{\infty} \frac{1}{n} \sum_{k=[n/2]}^{n} b_k{}^2$$

$$\le 2 \sum_{k=1}^{\infty} b_k{}^2 \sum_{n=k}^{2k+1} \frac{1}{n} \le 6 \sum_{k=1}^{\infty} b_k{}^2 < \infty.$$

This establishes (17) and completes the proof of the theorem.

As a first illustration, let $\lambda_n = n^{-1/2}$. Then (16) is satisfied, so

$$\sum_{n=1}^{\infty} \frac{|a_n|^2}{n} < \infty$$

for every $f \in H^1$. However, this follows from Hardy's inequality and the fact that $a_n \to 0$.

There is another application which is more interesting. Let $n_1, n_2, \ldots$ be a lacunary sequence of integers in the sense that

$$n_{k+1}/n_k \ge Q > 1.$$

Let

$$\lambda_n = \begin{cases} 1 & \text{if} \quad n = n_k \\ 0 & \text{otherwise.} \end{cases}$$

If $n_k \le N < n_{k+1}$,

$$\sum_{n=1}^{N} n^2 |\lambda_n|^2 = \sum_{j=1}^{k} n_j{}^2 \le n_k{}^2 \{1 + Q^{-2} + \cdots + Q^{2(1-k)}\} \le CN^2.$$

We thus obtain

PALEY'S THEOREM. If

$$f(z) = \sum_{n=0}^{\infty} a_n z^n \in H^1,$$

then for every lacunary sequence $\{n_k\}$,

$$\sum_{k=1}^{\infty} |a_{n_k}|^2 < \infty.$$

As a final corollary of Theorem 6.7, we now show that, more generally, $\{\lambda_n\}$ is a multiplier of H^1 into ℓ^q ($2 \leq q < \infty$) if (and only if) it satisfies (12). Let $\mu_n = |\lambda_n|^{q/2}$, and observe that (12) gives a condition equivalent to

$$\sum_{n=1}^{N} n^2 \mu_n^2 = O(N^2).$$

Thus $\{\mu_n\}$ multiplies H^1 into ℓ^2, which implies (since the coefficients of an H^1 function are bounded) that $\{\lambda_n\}$ multiplies H^1 into ℓ^q.

The multipliers from H^1 to ℓ^1 are more difficult to describe. Hardy's inequality shows that the sequence $\{(n + 1)^{-1}\}$ is one example. It is possible to characterize all the multipliers, but only by a condition difficult to verify in most situations. Thus the following theorem, interesting though it may be, is really more a translation of the problem than a solution.

THEOREM 6.8. The sequence $\{\lambda_n\}$ is a multiplier of H^1 into ℓ^1 if and only if there is a function $\psi \in L^\infty$ such that

$$|\lambda_n| = \frac{1}{2\pi} \int_0^{2\pi} e^{-int}\psi(t) \, dt, \qquad n = 0, 1, 2, \dots . \tag{18}$$

PROOF. The sufficiency of the condition (18) was established in Theorem 3.15. Conversely, suppose $\{\lambda_n\}$ is a multiplier from H^1 to ℓ^1. Then by the closed graph theorem, the mapping

$$\sum a_n z^n \to \{\lambda_n a_n\}$$

is a bounded operator from H^1 to ℓ^1:

$$\sum_{n=0}^{\infty} |\lambda_n a_n| \leq C\|f\|_1, \qquad f(z) = \sum a_n z^n.$$

In particular,

$$\phi(f) = \sum_{n=0}^{\infty} |\lambda_n| a_n$$

is a bounded linear functional on H^1. By the Hahn–Banach theorem, ϕ can be extended to a bounded linear functional Φ on L^1. Thus by the Riesz representation theorem, there exists $\psi \in L^\infty$ with

$$\Phi(f) = \frac{1}{2\pi} \int_0^{2\pi} f(e^{it})\overline{\psi(t)} \, dt, \qquad f \in L^1.$$

Choosing the H^1 function $f(z) = z^n$ $(n = 0, 1, 2, \ldots)$, we have

$$|\lambda_n| = \phi(f) = \Phi(f) = \frac{1}{2\pi} \int_0^{2\pi} e^{int} \overline{\psi(t)} \, dt,$$

which becomes (18) after conjugation.

EXERCISES

1. Show that Theorem 6.2 is best possible if $0 < p < 1$: For each positive sequence $\{k_n\}$ increasing to infinity, there exists $\sum a_n z^n \in H^p$ with

$$\sum k_n n^{p-2} |a_n|^p = \infty.$$

(This is also true if $1 \le p \le 2$; see Duren and Taylor [1].)

2. Show that Theorem 6.2 is false for $p > 2$.

3. Show that the converse to Theorem 6.2 is false $(0 < p \le 2)$.

4. Show that Paley's theorem does not generalize to Fourier series: There exists $f \in L^1$ with Fourier series $\sum c_n e^{in\theta}$, for which $\sum |c_{2^k}|^2 = \infty$. (*Suggestion*: Try $\sum (\log n)^{-a} \cos n\theta$, $a > 0$. See Theorem 4.5.)

5. Show that $f(z)$ may be analytic in $|z| < 1$ and continuous in $|z| \le 1$, yet $f'(z)$ have a radial limit almost nowhere. (This disproves the "Bloch–Nevanlinna conjecture" that $f \in N$ implies $f' \in N$. See Duren [4].)

6. A function $f(z)$ analytic in $|z| < 1$ is said to have finite Dirichlet integral $(f \in D)$ if

$$\frac{1}{\pi} \iint_{|z| < 1} |f'(z)|^2 \, dx \, dy = \sum_{n=1}^{\infty} n |a_n|^2 < \infty.$$

Show that $f' \in H^1$ implies $f \in D$, but that f need not be in D if $f' \in H^p$ for all $p < 1$.

7. Show that D is contained in H^p for all $p < \infty$, but D is not contained in H^∞.

NOTES

For the Hausdorff–Young theorem, see Zygmund [4]. Hardy and Little-wood [1] gave a long and difficult proof of Theorem 6.2 in their original paper. The idea to prove it for $0 < p < 1$ via Theorem 5.11 is due to Flett [2]. The fact that $a_n = o(n^{1/p-1})$ for $f \in H^p$ $(p < 1)$ is due to Hardy and Littlewood [5], although in the Soviet literature it is often ascribed to G. A. Fridman, who rediscovered it in 1949. Evgrafov [1] gave a direct construction to show that this estimate is best possible; a simpler argument was given by Duren and Taylor [1]. Theorem 6.5 (and its corollary) and Theorem 6.6 are due to Duren and Shields [1, 2]. Hardy and Littlewood [7, 8] gave a sufficient condition for multipliers from H^p to H^q ($1 \le p \le 2 \le q < \infty$) which includes the "if" part of Theorem 6.7; the proof in the text is theirs. They also stated without proof a sufficient condition for multipliers from H^p to H^q ($0 < p < 1 \le q < \infty$). A proof appears in Duren and Shields [2]. The closed graph theorem has often been applied to multiplier problems in various spaces. J. H. Wells pointed out to the author that the necessity of (16) could be proved by this method. Paley [1] proved "Paley's theorem" and inspired the original Hardy–Littlewood work on multipliers. A converse to Paley's theorem, which Rudin [4] proved by explicit construction, is a simple consequence of the necessity of the multiplier condition (16); see Stein and Zygmund [1]. Hedlund [1] recently gave a sufficient condition for multipliers from H^p ($1 < p < 2$) to H^2 which extends that of Hardy and Littlewood. The condition (12) is necessary but *not* sufficient if $p = 1$ and $q < 2$; see Duren and Shields [2]. E. M. Stein [1] has given a sufficient condition for multipliers from H^p to H^p ($0 < p < 1$). For further information on multipliers, see Kaczmarz [1], Caveny [1, 2], Wells [1], Duren and Shields [1, 2], and Duren [3].

The coefficients of inner functions have been studied by Newman and Shapiro [1]. They show that no inner function except a finite Blaschke product can have coefficients $o(1/n)$, although the coefficients of an infinite Blaschke product can be $O(1/n)$. The coefficients of a singular inner function cannot be $o(n^{-3/4})$, but they can be $O(n^{-3/4})$.

Information is also available on the coefficients of functions of bounded characteristic. Mergelyan has shown that if $f(z) = \sum a_n z^n \in N$, then

$$\limsup_{n \to \infty} n^{-1/2} \log|a_n| \le C,$$

where C can be given explicitly in terms of f. (See Privalov [4].) Cantor [1] recently proved that if $f \in N$ and A_n denotes the matrix (a_{i+j}), $0 \le i, j \le n$, then $|\det A_n|^{1/n} \to 0$ as $n \to \infty$.

In order to solve problems concerning H^p functions, it is often advantageous to view H^p as a linear space and to use the methods of functional analysis. We have already seen in Chapter 6 that the closed graph theorem is an effective tool for describing coefficient multipliers. The same tool will be used in Chapter 9 to discuss interpolation problems. In the present chapter, we shall study the linear space structure of H^p in some detail. One major objective is to represent the continuous linear functionals on H^p both for $p \geq 1$ and in the more interesting case $p < 1$. The results for $p \geq 1$ help to solve an approximation problem (Section 7.3) and prepare the ground for a full discussion of extremal and interpolation problems in Chapters 8 and 9. The dual space structure of H^p with $p < 1$ is applied to demonstrate the failure of the Hahn-Banach theorem in a non-locally convex space with "reasonable" properties. This chapter concludes with a description of the extreme points of the unit sphere in H^1, a topic also of interest in the general context of linear space theory.

7.1. QUOTIENT SPACES AND ANNIHILATORS

Let us begin by recalling a few general Banach space concepts. Let X be a Banach space, and let S be a (*closed*) subspace. A *coset of X modulo S* is a subset $\xi = x + S$ consisting of all elements of the form $x + y$, where x is some fixed member of X and y ranges over S. Two cosets are either identical or disjoint. The *quotient space* X/S has as its elements all distinct cosets of X modulo S. With the natural definitions of addition and scalar multiplication, X/S is a linear space. To be specific,

$$(x_1 + S) + (x_2 + S) = (x_1 + x_2) + S$$

and

$$\alpha(x + S) = \alpha x + S.$$

The zero element of X/S is the coset S. Finally, the norm of a coset $\xi = x + S$ is defined by

$$\|\xi\| = \inf_{y \in S} \|x + y\|.$$

It is easy to check that this is a genuine norm. Since S is closed, $\|\xi\| = 0$ implies $\xi = S$. The relations $\|\xi\| \geq 0$, $\|\alpha\xi\| = |\alpha| \, \|\xi\|$, and $\|\xi_1 + \xi_2\| \leq \|\xi_1\| + \|\xi_2\|$ are obvious.

Under the given norm, X/S is complete, and therefore is itself a Banach space. To see this, let $\{\xi_n\}$ be a Cauchy sequence of cosets. Choose a sequence of integers $0 < n_1 < n_2 < \cdots$ such that

$$\|\xi_{n_k} - \xi_{n_{k+1}}\| \leq 2^{-k}, \qquad k = 1, 2, \ldots.$$

Then choose $x_k \in \xi_{n_k}$ ($k = 1, 2, \ldots$) such that

$$\|x_k - x_{k+1}\| < 2\|\xi_{n_k} - \xi_{n_{k+1}}\|.$$

With this construction, $\{x_k\}$ is a Cauchy sequence, so x_k tends to a limit $x \in X$. Let $\xi = x + S$. Then $\xi_{n_k} \to \xi$, because (by definition of the norm)

$$\|\xi_{n_k} - \xi\| \leq \|x_k - x\|.$$

Finally, since $\{\xi_n\}$ is a Cauchy sequence, this implies $\xi_n \to \xi$.

The *annihilator* of the subspace S is the set $S^\perp$ of all linear functionals $\phi \in X^*$ such that $\phi(x) = 0$ for all $x \in S$. It can be easily verified that $S^\perp$ is a subspace of X^*. The following results play an essential role in the theory of extremal problems (Chapter 8).

THEOREM 7.1. The quotient space $X^*/S^\perp$ is isometrically isomorphic to S^*. Furthermore, for each fixed $\phi \in X^*$,

$$\sup_{x \in S, \, \|x\| \leq 1} |\phi(x)| = \min_{\psi \in S^\perp} \|\phi + \psi\|,$$

where "min" indicates that the infimum is attained.

THEOREM 7.2. The space $(X/S)^*$ is isometrically isomorphic to $S^\perp$. Furthermore, for each fixed $x \in X$,

$$\max_{\psi \in S^\perp, \, \|\psi\| \leq 1} |\psi(x)| = \inf_{y \in S} \|x + y\|,$$

where "max" indicates that the supremum is attained.

PROOF OF THEOREM 7.1. For each fixed $\psi \in S^*$, the class of all extensions $\phi \in X^*$ is a coset in $X^*/S^\perp$. It is clear that this correspondence between S^* and $X^*/S^\perp$ is an isomorphism. It is also an isometry. In fact, $\|\psi\| \leq \|\phi\|$ for every extension ϕ; and, by the Hahn–Banach theorem, there is at least one extension for which $\|\psi\| = \|\phi\|$. In other words, for the coset of extensions of ψ, the infimum defining the norm is *attained* and is equal to $\|\psi\|$.

PROOF OF THEOREM 7.2. In terms of a given $\Phi \in (X/S)^*$, one can define a linear functional ϕ in $S^\perp$ unambiguously by

$$\phi(x) = \Phi(x + S).$$

Conversely, given $\phi \in S^\perp$, this relation defines a linear functional Φ on X/S. The correspondence is clearly an isomorphism. We assert that the boundedness of either functional implies that of the other, and in fact $\|\phi\| = \|\Phi\|$. This follows from the relations

$$|\phi(x)| = |\Phi(x + S)| \leq \|\Phi\| \, \|x + S\| \leq \|\Phi\| \, \|x\|;$$
$$|\Phi(x + S)| = |\phi(x)| = |\phi(x + y)| \leq \|\phi\| \, \|x + y\|, \qquad y \in S.$$

Hence $S^\perp$ and $(X/S)^*$ are isometrically isomorphic.

To prove the rest of the theorem, observe first that for any $\phi \in S^\perp$ with $\|\phi\| \leq 1$, and for any $x \in X$ and $y \in S$,

$$|\phi(x)| = |\phi(x + y)| \leq \|x + y\|.$$

Thus

$$\sup_{\phi \in S^\perp, \, \|\phi\| \leq 1} |\phi(x)| \leq \inf_{y \in S} \|x + y\|. \tag{1}$$

On the other hand, given $x \in X$, a corollary of the Hahn–Banach theorem (see Dunford and Schwartz [1], p. 65) shows the existence of $\Phi \in (X/S)^*$ such that

$$|\Phi(x + S)| = \|x + S\| \qquad \text{and} \qquad \|\Phi\| = 1.$$

Now let $\phi \in S^\perp$ correspond to Φ as above, so that $\phi(x) = \Phi(x + S)$ and $\|\phi\| = \|\Phi\| = 1$. Equality therefore holds in (1), and the supremum is attained.

7.2. REPRESENTATION OF LINEAR FUNCTIONALS

As we saw in Chapter 3, H^p is a Banach space if $1 \leq p \leq \infty$, with norm $\|f\| = M_p(1, f)$. The polynomials are dense in H^p if $0 < p < \infty$. If $1 \leq p \leq \infty$, the set of boundary functions of H^p is the subspace of L^p for which

$$\int_0^{2\pi} e^{in\theta} f(e^{i\theta}) \, d\theta = 0, \qquad n = 1, 2, \dots.$$

In particular, if each $f \in H^p$ is identified with its boundary function, H^p can be regarded as a subspace of L^p, $0 < p \leq \infty$.

According to the Riesz representation theorem, every bounded linear functional ϕ on L^p ($1 \leq p < \infty$) has a unique representation

$$\phi(f) = \frac{1}{2\pi} \int_0^{2\pi} f(e^{i\theta}) g(e^{i\theta}) \, d\theta, \qquad g \in L^q, \tag{2}$$

where $1/p + 1/q = 1$. In fact, $\|\phi\| = \|g\|_q$, and $(L^p)^*$ is isometrically isomorphic to L^q. Since H^p is a subspace of L^p, then, Theorem 7.1 can be used to describe $(H^p)^*$ if the annihilator of H^p in $(L^p)^*$ can be determined. But if $g \in L^q$ annihilates every H^p function, then surely

$$\int_0^{2\pi} e^{in\theta} g(e^{i\theta}) \, d\theta = 0, \qquad n = 0, 1, 2, \dots.$$

Therefore $g(e^{i\theta})$ is the boundary function of some $g(z) \in H^q$, and $g(0) = 0$. Call this class of functions H_0^q. Conversely, if $g \in H_0^q$, it is clear that

$$\int_0^{2\pi} f(e^{i\theta}) g(e^{i\theta}) \, d\theta = 0$$

for every $f \in H^p$. Hence H_0^q is the annihilator of H^p, and it follows from Theorem 7.1 that $(H^p)^*$ is isometrically isomorphic to L^q/H_0^q. Actually, we may as well replace L^q/H_0^q by L^q/H^q, since the correspondence $\xi \leftrightarrow e^{i\theta}\xi$ between cosets of the two spaces is an isometric isomorphism.

It is even possible to give a canonical representation of the bounded linear functionals on H^p. Any $\phi \in (H^p)^*$ can be extended (by the Hahn–Banach theorem) to a functional on L^p, and hence may be represented in the form (2) for some $g \in L^q$. This representation is certainly not unique. Two functions g_1 and g_2 belonging to the same coset of L^q/H_0^q, so that

$$g_1(e^{i\theta}) - g_2(e^{i\theta}) \sim \sum_{n=1}^{\infty} c_n e^{in\theta},$$

obviously represent the same functional ϕ on H^p. Functions in different cosets, however, generate different functionals. Thus for $p > 1$, the representation becomes unique if we distinguish in each coset that function g for which

$$\int_0^{2\pi} e^{-in\theta} g(e^{i\theta})\, d\theta = 0, \qquad n = 1, 2, \ldots.$$

Equivalently, there is a unique function $g \in H^q$ for which

$$\phi(f) = \frac{1}{2\pi} \int_0^{2\pi} f(e^{i\theta}) \overline{g(e^{i\theta})}\, d\theta \tag{3}$$

for all $f \in H^p$, $1 < p < \infty$. Since $1 < q < \infty$, the M. Riesz theorem (Theorem 4.1) guarantees that the "analytic projection" g of the original L^q function is in H^q (see Chapter 4, Exercise 1). In summary:

THEOREM 7.3. For $1 \leq p < \infty$, the space $(H^p)^*$ is isometrically isomorphic to L^q/H^q, where $1/p + 1/q = 1$. Furthermore, if $1 < p < \infty$, each $\phi \in (H^p)^*$ is representable in the form (3) by a unique function $g \in H^q$, while each $\phi \in (H^1)^*$ can be represented in the form (3) by some $g \in L^\infty$.

We might have chosen to put $g(e^{-i\theta})$ instead of $\overline{g(e^{i\theta})}$ in (3). This would have the advantage of setting up an isomorphism between $(H^p)^*$ and H^q. But in either case the correspondence need not be an isometry. In fact, $\|\phi\|$ is equal to the norm of the coset determined by $\overline{g(e^{i\theta})}$ (or by $g(e^{-i\theta})$), so that only the inequality

$$\|\phi\| \leq \|g\| \leq A_p \|\phi\|, \qquad 1 < p < \infty,$$

is true generally. The right-hand inequality comes from the M. Riesz theorem; A_p is a constant independent of ϕ. In the case $p = 2$ it is easy to see that $\|\phi\| = \|g\|$.

7.3. BEURLING'S APPROXIMATION THEOREM

The representation of linear functionals can be applied to prove an interesting theorem on polynomial approximation in H^p, $1 \leq p < \infty$. Recall that a function $f \in H^p$ has a canonical factorization

$$f(z) = B(z)S(z)F(z),$$

where $B(z)$ is a Blaschke product, $S(z)$ is a singular inner function generated by a nondecreasing singular function $\mu(t)$, and $F(z)$ is an outer function. The inner function $f_0(z) = B(z)S(z)$ is called the *inner factor* of $f(z)$. Let $g_0(z)$ be another inner function, with an associated singular function $\nu(t)$. f_0 is said to be a *divisor* of g_0 if $g_0(z)/f_0(z)$ is an inner function. This is clearly the case if and only if every zero of $f_0(z)$ in $|z| < 1$ is also a zero of $g_0(z)$ (with the same or higher multiplicity) and $[\nu(t) - \mu(t)]$ is nondecreasing.

For fixed $f \in H^p$, let $\mathscr{P}[f]$ denote the subspace generated by the functions $z^n f(z)$, $n = 0, 1, 2, \ldots$. Thus $\mathscr{P}[f]$ consists of all H^p functions which can be approximated by polynomial multiples of f. The problem is to characterize $\mathscr{P}[f]$. It is already known that $\mathscr{P}[1] = H^p$, since the polynomials are dense in H^p. More generally, $\mathscr{P}[f]$ can be described as follows.

THEOREM 7.4 (Beurling). Let f and g be H^p functions $(1 \le p < \infty)$, not identically zero, with inner factors f_0 and g_0, respectively. Then $g \in \mathscr{P}[f]$ if and only if f_0 is a divisor of g_0.

COROLLARY 1. $\mathscr{P}[f] = \mathscr{P}[f_0]$. If $\mathscr{P}[f_0] = \mathscr{P}[g_0]$, then $f_0 = g_0$.

PROOF OF THEOREM. Under the assumption that f_0 divides g_0, it is to be shown that $g \in \mathscr{P}[f]$. Since

$$\|Pf - g\| = \|PF - (g_0/f_0)G\|,$$

it suffices to prove that $\mathscr{P}[F] = H^p$ for every outer function $F(z)$. If $\mathscr{P}[F]$ is not the whole space, there is a nontrivial bounded linear functional which annihilates it. That is, there exists a function $h(e^{i\theta}) \in L^q$ $(1/p + 1/q = 1)$ such that $\overline{h(e^{i\theta})} \notin H_0^q$ and

$$\int_0^{2\pi} e^{in\theta} F(e^{i\theta}) \overline{h(e^{i\theta})} \, d\theta = 0, \qquad n = 0, 1, 2, \ldots.$$

Thus $F(e^{i\theta}) \overline{h(e^{i\theta})} = k(e^{i\theta})$, where $k(z) \in H_0^1$. Since $k(e^{i\theta})/F(e^{i\theta}) \in L^q$, it follows from Theorem 2.11 that

$$\overline{h(e^{i\theta})} = \frac{k(e^{i\theta})}{F(e^{i\theta})} \in H_0^q,$$

a contradiction.

Conversely, if $g \in \mathscr{P}[f]$ it is obvious that g must vanish in $|z| < 1$ wherever f does, since norm convergence implies pointwise convergence. Thus $h(z) = g_0(z)/f_0(z)$ is analytic. It has to be proved that h is an inner function. But since $g \in \mathscr{P}[f]$, there is a sequence of polynomials $\{P_n\}$ for which $\|P_n f - g\| \to 0$. Division by $f_0(e^{i\theta})$ then shows that $h(e^{i\theta})G(e^{i\theta})$ is an L^p function approximable in norm by H^p functions. Thus $hG \in H^p$, which implies that $h \in H^p$. This shows (by Theorem 2.11) that h is an inner function, completing the proof.

COROLLARY 2. Let $0 < p < \infty$. Then for any inner function f_0, $\mathscr{P}[f_0] = f_0 \cdot H^p$.

This is true even for $0 < p < 1$, since the proof used only the fact that H^p is the L^p closure of the polynomials.

7.4. LINEAR FUNCTIONALS ON H^p, $0 < p < 1$

Although H^p is not normable in the case $p < 1$, its bounded linear functionals can still be defined in the usual manner. Thus a linear functional ϕ on H^p is said to be bounded if

$$\|\phi\| = \sup_{\|f\|_p = 1} |\phi(f)| < \infty.$$

It is easy to verify that a linear functional is bounded if and only if it is continuous. It can also be checked that the bounded linear functionals on H^p form a Banach space under the given norm.

We observed in Section 6.4 that H^p is an F-space even if $p < 1$. Hence the principle of uniform boundedness (Banach–Steinhaus theorem) still applies: every pointwise bounded sequence of bounded linear functionals on H^p is uniformly bounded. (See Dunford and Schwartz [1], Chapter II.)

We are about to derive a complete representation of the bounded linear functionals on H^p, $p < 1$. First, however, we need to introduce some notation. Let A denote the class of functions analytic in $|z| < 1$ and continuous in $|z| \leq 1$. It is convenient to write $f \in \Lambda_\alpha$ to indicate that $f \in A$ and its boundary function $f(e^{i\theta})$ belongs to the Lipschitz class Λ_α, $0 < \alpha \leq 1$. Similarly, $f \in \Lambda_*$ will mean that $f \in A$ and $f(e^{i\theta}) \in \Lambda_*$. (See Section 5.1.)

THEOREM 7.5. To each bounded linear functional ϕ on H^p, $0 < p < 1$, there corresponds a unique function $g \in A$ such that

$$\phi(f) = \lim_{r \to 1} \frac{1}{2\pi} \int_0^{2\pi} f(re^{i\theta}) g(e^{-i\theta}) \, d\theta, \qquad f \in H^p. \tag{4}$$

If $(n+1)^{-1} < p < n^{-1}$ $(n = 1, 2, \ldots)$, then $g^{(n-1)} \in \Lambda_\alpha$, where $\alpha = 1/p - n$. Conversely, for any g with $g^{(n-1)} \in \Lambda_\alpha$, the limit (4) exists for all $f \in H^p$ and defines a bounded linear functional. If $p = (n+1)^{-1}$, then $g^{(n-1)} \in \Lambda_*$; and conversely, any g with $g^{(n-1)} \in \Lambda_*$ defines through (4) a bounded linear functional on H^p.

PROOF. Given $\phi \in (H^p)^*$, let $b_k = \phi(z^k)$, $k = 0, 1, \ldots$. Then

$$|b_k| \leq \|\phi\| \, \|z^k\| = \|\phi\|,$$

so the function

$$g(z) = \sum_{k=0}^{\infty} b_k z^k$$

is well defined and analytic in $|z| < 1$. Suppose now that

$$f(z) = \sum_{k=0}^{\infty} a_k z^k \in H^p.$$

For fixed ρ, $0 < \rho < 1$, let $f_\rho(z) = f(\rho z)$. Since f_ρ is the uniform limit on $|z| = 1$ of the partial sums of its power series, and since ϕ is continuous, it follows that

$$\phi(f_\rho) = \lim_{N \to \infty} \phi\left(\sum_{k=0}^{N} a_k \rho^k z^k\right) = \sum_{k=0}^{\infty} a_k b_k \rho^k.$$

But $f_\rho \to f$ in H^p norm as $\rho \to 1$, so

$$\phi(f) = \lim_{\rho \to 1} \sum_{k=0}^{\infty} a_k b_k \rho^k. \tag{5}$$

To deduce (4) from (5), it would suffice to show $g \in H^1$. But for fixed ζ, $|\zeta| < 1$, let

$$f(z) = (1 - \zeta z)^{-1} = \sum_{k=0}^{\infty} \zeta^k z^k.$$

Then by (5),

$$\phi(f) = \sum_{k=0}^{\infty} \zeta^k b_k = g(\zeta).$$

Hence

$$|g(\zeta)| \le \|\phi\| \, \|f\|_p \le \|\phi\| \, \|(1-z)^{-1}\|_p \, ;$$

that is, $g \in H^\infty$. This proves (4), with $g \in H^\infty$.

Now suppose $(n+1)^{-1} < p < n^{-1}$, and let

$$F(z) = \frac{n! \, z^n}{(1 - \zeta z)^{n+1}}, \qquad |\zeta| < 1.$$

In view of (4) and the Cauchy formula, we have

$$\phi(F) = g^{(n)}(\zeta).$$

It now follows from the lemma in Section 4.6 that

$$|g^{(n)}(\zeta)| \le \|\phi\| \, \|F\|_p = O((1 - |\zeta|)^{1/p - n - 1}),$$

so that $g^{(n-1)} \in \Lambda_\alpha$ with $\alpha = 1/p - n$, by Theorem 5.1. If $p = (n+1)^{-1}$, a similar argument shows $g^{(n+1)}(\zeta) = O((1 - |\zeta|)^{-1})$, which implies $g^{(n-1)} \in \Lambda_*$, by Theorem 5.3.

To prove the converse, we first suppose $(n+1)^{-1} < p < n^{-1}$ and that $g(z) = \sum b_k z^k$ is given with $g^{(n-1)} \in \Lambda_\alpha$, $\alpha = 1/p - n$. It is to be shown that for every $f(z) = \sum a_k z^k$ in H^p, the function

$$\psi(r) = \sum_{k=0}^{\infty} a_k b_k r^k$$

has a limit as $r \to 1$, and that

$$\lim_{r \to 1} |\psi(r)| \le C\|f\|_p.$$

We shall prove the existence of the limit by showing that

$$\int_0^1 |\psi'(r)| \, dr < \infty. \tag{6}$$

Setting

$$h(z) = z^{n-1}g(z)$$

and

$$f_{\nu+1}(z) = \int_0^z f_\nu(\zeta) \, d\zeta, \qquad \nu = 0, 1, \ldots,$$

where $f_0(z) = f(z)$, we have the relation

$$r^n\psi'(r^2) = \frac{1}{2\pi} \int_0^{2\pi} e^{-in\theta} f_{n-1}(re^{i\theta}) h^{(n)}(re^{-i\theta}) \, d\theta, \tag{7}$$

which can be checked by comparing power series. By Theorem 5.12,

$$f_\nu \in H^{p_\nu}, \qquad \nu = 1, 2, \ldots, n,$$

where $p_0 = p$ and

$$p_{\nu+1} = \frac{p_\nu}{1 - p_\nu}.$$

Simple induction shows that

$$p_\nu = \frac{p}{1 - \nu p}, \qquad \nu = 1, 2, \ldots, n.$$

Hence

$$p < p_1 < \cdots < p_{n-1} < 1 < p_n < \infty.$$

On the other hand, the fact that $g^{(n-1)} \in \Lambda_\alpha$ gives

$$|h^{(n)}(re^{i\theta})| \leq \frac{C}{(1-r)^{1-\alpha}},$$

by Theorem 5.1. It therefore follows from (7) that

$$r^n|\psi'(r^2)| \leq C(1-r)^{\alpha-1} M_1(r, f_{n-1}), \qquad \alpha = 1/p - n.$$

But according to Theorem 5.11,

$$\int_0^1 (1-r)^\beta M_1(r, f_{n-1}) \, dr < \infty,$$

where

$$\beta = \frac{1}{p_{n-1}} - 2 = \alpha - 1.$$

This proves (6) for $(n+1)^{-1} < p < n^{-1}$.

If $p = (n + 1)^{-1}$ and $g^{(n-1)} \in \Lambda_*$, (6) can be established by a similar argument. We omit the details, since they involve the theory of fractional integration, and a development of this background material would carry us too far afield. A full account may be found in Duren, Romberg, and Shields [1].

To complete the proof, we now show that if $g(z) = \sum b_k z^k$ is any function such that

$$\phi(f) = \lim_{r \to 1} \sum_{k=0}^{\infty} a_k b_k r^k$$

exists for every $f(z) = \sum a_k z^k$ in H^p, then $\phi \in (H^p)^*$. For fixed $r < 1$, let

$$\phi_r(f) = \sum_{k=0}^{\infty} a_k b_k r^k.$$

By Theorem 6.5, $\phi_r \in (H^p)^*$ for each r. But for each fixed $f \in H^p$,

$$\sup_r |\phi_r(f)| < \infty.$$

Thus by the principle of uniform boundedness,

$$\sup_r \|\phi_r\| < \infty,$$

which implies $\phi \in (H^p)^*$. This concludes the proof.

We remark that the principle of uniform boundedness need not have been used. The proofs of the various auxiliary theorems actually give norm estimates which could have been carried through the proof to establish the boundedness of ϕ directly.

The function $g(z) = (1 - \zeta z)^{-1}$, $|\zeta| < 1$, provides an interesting example. Here the integral in (4) reduces simply to $f(r\zeta)$, so that $\phi(f) = f(\zeta)$. Point evaluation is therefore a bounded linear functional on H^p, $p < 1$. (This can also be proved directly; see the lemma in Section 3.2.) In particular, there are enough linear functionals on H^p to distinguish elements of the space. This contrasts sharply with the situation for L^p ($p < 1$), where there are no bounded linear functionals at all except the zero functional.

7.5. FAILURE OF THE HAHN–BANACH THEOREM

We have just seen that H^p has enough continuous linear functionals to distinguish elements of the space, even if $p < 1$. On the other hand, we shall now show that there are not always enough functionals to separate points from subspaces. That is, given a proper subspace M of H^p ($p < 1$) and an H^p function $f \notin M$, there may not be any functional $\phi \in (H^p)^*$ such that $\phi(M) = 0$

and $\phi(f) \neq 0$. To show this, we shall construct a proper subspace of H^p which is annihilated by no functional $\phi \in (H^p)^*$ except the zero functional.

Before turning to the actual construction, we have to develop some background material which is of interest itself. Let

$$S(z) = \exp\left\{ - \int_0^{2\pi} \frac{e^{it} + z}{e^{it} - z} \, d\mu(t) \right\}$$

be a singular inner function; and let $\omega(t; \mu)$ denote the modulus of continuity of μ.

LEMMA 1. If $\omega(t; \mu) = O(t \log 1/t)$, then there exist positive constants a and C such that

$$|S(z)| \geq C(1 - r)^a, \qquad |z| < 1.$$

PROOF. We have

$$- \log|S(z)| = \int_{\theta - \pi}^{\theta + \pi} P(r, t - \theta) \, d\mu(t)$$

$$= P(r, \pi)\{\mu(\theta + \pi) - \mu(\theta - \pi)\}$$

$$- \int_0^{\pi} \frac{\partial P}{\partial t} (r, t)\{\mu(\theta + t) - \mu(\theta - t)\} \, dt$$

$$\leq k_1 + \int_0^{\pi} \left(-\frac{\partial P}{\partial t} \right) |\mu(\theta + t) - \mu(\theta - t)| \, dt$$

$$\leq k_2 + k_3 \int_0^{1/e} \left(-\frac{\partial P}{\partial t} \right) t \log \frac{1}{t} \, dt.$$

But another integration by parts shows this last integral is less than

$$k_4 + \int_0^{1/e} P(r, t) \log \frac{1}{t} \, dt$$

$$\leq k_4 + \log \frac{1}{1 - r} + \int_0^{1-r} P(r, t) \log \frac{1}{t} \, dt$$

$$\leq k_4 + \log \frac{1}{1 - r} + \frac{2}{1 - r} \int_0^{1-r} \log \frac{1}{t} \, dt$$

$$= (k_4 + 2) + 3 \log \frac{1}{1 - r}.$$

Hence

$$- \log|S(z)| \leq k + a \log \frac{1}{1-r},$$

which proves the lemma.

We now introduce the space B_α ($\alpha > 0$) of functions $f(z) = \sum a_n z^n$ analytic in $|z| < 1$, such that

$$\|f\|_\alpha^2 = \sum_{n=0}^{\infty} (n+1)^{-\alpha} |a_n|^2 < \infty.$$

B_α is a Hilbert space with inner product

$$(f, g) = \sum_{n=0}^{\infty} (n+1)^{-\alpha} a_n \overline{b_n},$$

where $g(z) = \sum b_n z^n$. Obviously, the polynomials are dense in B_α, for each $\alpha > 0$. A straightforward calculation shows that

$$C_1 \|f\|_\alpha^2 \leq \int_0^{2\pi} \int_0^1 (1-r)^{\alpha-1} |f(re^{i\theta})|^2 \, dr \, d\theta$$

$$\leq C_2 \|f\|_\alpha^2, \tag{8}$$

for some positive constants C_1 and C_2 depending only on α.

A function $f \in B_\alpha$ will be called an α-*outer function* if the set of all polynomial multiples Pf is dense in B_α. Clearly, f is an α-outer function if and only if 1 belongs to the B_α closure of the polynomial multiples of f.

LEMMA 2. If $\omega(t; \mu) = O(t \log 1/t)$, then S is an α-outer function, for every $\alpha > 0$.

PROOF. By Lemma 1 and (8), $S^{-1/m} \in B_\alpha$ for some positive integer m. Given $\varepsilon > 0$, it follows from (8) that

$$\|PS^{1/m} - 1\|_\alpha \leq C\|P - S^{-1/m}\|_\alpha < \varepsilon$$

for a suitable polynomial P. This shows that $S^{1/m}$ is an α-outer function. To conclude that S is an α-outer function, we need only observe that if $f \in H^\infty$ and $g \in B_\alpha$ are α-outer functions, so is their product fg. Indeed, by (8),

$$\|PQfg - 1\|_\alpha \leq \|Pf(Qg - 1)\|_\alpha + \|Pf - 1\|_\alpha$$
$$\leq C\|Pf\|_\infty \|Qg - 1\|_\alpha + \|Pf - 1\|_\alpha.$$

Choose P to make the second term small; then choose Q to make the first term small. This shows that fg is an α-outer function, and the proof is complete.

We now consider the Hilbert space H^2 and denote by $M^\perp$ the orthogonal complement of a subspace M of H^2. For fixed $f \in H^\infty$, fH^p will denote the subspace of H^p $(0 < p \le \infty)$ consisting of all multiples fg, $g \in H^p$.

THEOREM 7.6. Let S be a singular inner function such that $\omega(t; \mu) = O(t \log 1/t)$. Then $(SH^2)^\perp$ contains no nonnull function $g(z) = \sum b_n z^n$ such that

$$\sum_{n=1}^{\infty} n^\gamma |b_n|^2 < \infty, \qquad \gamma > 0. \tag{9}$$

PROOF. Let $S(z) = \sum a_n z^n$. Suppose $g \in H^2$ is orthogonal to SH^2. Then g is orthogonal to $z^k S(z)$, $k = 0, 1, \ldots$; so

$$\sum_{n=k}^{\infty} a_{n-k} \overline{b_n} = 0, \qquad k = 0, 1, \ldots. \tag{10}$$

Now define

$$c_n = (n+1)^\gamma b_n.$$

If (9) is satisfied, then

$$h(z) = \sum_{n=0}^{\infty} c_n z^n \in B_\gamma;$$

while the conditions (10) are equivalent to

$$\sum_{n=k}^{\infty} \frac{a_{n-k} \overline{c_n}}{(n+1)^\gamma} = 0, \qquad k = 0, 1, \ldots.$$

This says that h is orthogonal to $z^k S(z)$ $(k = 0, 1, \ldots)$ *in the space* B_γ. But by Lemma 2, the functions $z^k S(z)$ span B_γ. Hence $h = 0$, which implies $g = 0$.

One final lemma is needed.

LEMMA 3. If $g(z) = \sum b_n z^n \in \Lambda_\alpha$ for some $\alpha > 0$, then

$$\sum_{n=1}^{\infty} n^\gamma |b_n|^2 < \infty \qquad \text{for all } \gamma < 2\alpha.$$

PROOF. By Theorem 5.1,

$$|g'(z)| \le C(1-r)^{\alpha-1}.$$

Thus

$$\sum_{n=1}^{\infty} n^2 r^{2n-2} |b_n|^2 = \frac{1}{2\pi} \int_0^{2\pi} |g'(re^{i\theta})|^2 \, d\theta \le C^2 (1-r)^{2\alpha-2}.$$

It follows that

$$r^{2N} \sum_{n=1}^{N} n^2 |b_n|^2 \le C^2 (1-r)^{2\alpha-2}.$$

Setting $r = 1 - 1/N$, $N = 2, 3, \ldots$, we find

$$s_N = \sum_{n=1}^{N} n^2 |b_n|^2 \leq K N^{2-2\alpha}, \qquad K = 16C^2.$$

A summation by parts now gives

$$\sum_{n=1}^{N} n^\gamma |b_n|^2 = \sum_{n=1}^{N} n^2 |b_n|^2 \cdot n^{\gamma-2}$$

$$= \sum_{n=1}^{N} s_n [n^{\gamma-2} - (n+1)^{\gamma-2}] + s_N (N+1)^{\gamma-2} = O(1)$$

if $\gamma < 2\alpha$.

We are now ready to give the construction promised at the beginning of this section.

THEOREM 7.7. Let S be a singular inner function with

$$\omega(t; \mu) = O(t \log 1/t).$$

Then for each p ($0 < p < 1$), the only bounded linear functional on H^p which annihilates the subspace SH^p is the zero functional.

PROOF. Suppose $\phi \in (H^p)^*$ annihilates SH^p. Then, in particular,

$$\phi(z^n S(z)) = \frac{1}{2\pi} \int_0^{2\pi} e^{in\theta} S(e^{i\theta}) \overline{g(e^{i\theta})} \, d\theta = 0, \qquad n = 0, 1, \ldots,$$

where $g(z) = \sum b_n z^n \in \Lambda_\alpha$ for some $\alpha > 0$ (see Theorem 7.5). This implies that g, regarded as an element of H^2, is orthogonal to the subspace SH^2. Therefore, by Theorem 7.6,

$$\sum_{n=1}^{\infty} n^\gamma |b_n|^2 = \infty \qquad \text{for each } \gamma > 0,$$

unless $g = 0$. But this contradicts Lemma 3, since $g \in \Lambda_\alpha$.

We remark that SH^p is a *proper* subspace unless $S = 1$.

COROLLARY 1. If S is a singular inner function as described in the theorem, the quotient space $H^p/(SH^p)$ has no continuous linear functionals except the zero functional, for each $p < 1$.

COROLLARY 2. If $p < 1$, there is a subspace M of H^p and a bounded linear functional on M which cannot be extended to a bounded linear functional on H^p.

The proofs are left as exercises.

7.6. EXTREME POINTS

A set S in a linear space X is said to be *convex* if whenever x_1 and x_2 are in S, every proper convex combination

$$ax_1 + (1 - a)x_2, \qquad 0 < a < 1,$$

is also in S. An element $x \in S$ is called an *extreme point* of S if it is not a proper convex combination of any two distinct points in S. There is considerable interest in finding the extreme points of a convex set, especially in view of the Krein–Milman theorem, which states that a compact convex set in a locally convex topological vector space is the closed convex hull of its extreme points. In particular, such a set *has* extreme points.

In any Banach space X, the problem arises to describe the extreme points of the unit sphere

$$S = \{x \in X : \|x\| \le 1\}.$$

Every extreme point obviously lies on the boundary of S; that is, it has unit norm. It is useful to observe that if a point x with $\|x\| = 1$ is not an extreme point, it can be represented as the midpoint of the segment joining two distinct points $(x + y)$ and $(x - y)$ in S. Furthermore,

$$\|x + y\| = \|x - y\| = 1, \tag{11}$$

since

$$1 = \|x\| = \tfrac{1}{2}\{\|x + y\| + \|x - y\|\} \le 1.$$

Hence a point x with $\|x\| = 1$ is an extreme point if and only if (11) implies $y = 0$.

If X is a Hilbert space, it is not hard to show that *every* boundary point of S is an extreme point. More generally, the same is true if X is a *uniformly convex* space; that is, if for each $\varepsilon > 0$ there exists a $\delta > 0$ such that $\|x\| = \|y\| = 1$ and $\|x - y\| > \varepsilon$ imply $\|(x + y)/2\| < 1 - \delta$. If $1 < p < \infty$, the space L^p, and therefore H^p, is uniformly convex, so every boundary point is an extreme point of the unit sphere. On the other hand, it is not hard to show that the unit sphere in L^1 has no extreme points at all. What is the situation for H^1?

THEOREM 7.8 (deLeeuw–Rudin). A function $f \in H^1$ with $\|f\| = 1$ is an extreme point of the unit sphere in H^1 if and only if f is an outer function.

PROOF. Suppose first that f is an outer function and $\|f\| = 1$. To show f is an extreme point, suppose $g \in H^1$ and

$$\|f + g\| = \|f - g\| = 1. \tag{12}$$

Let $h = g/f$. Then $h \in H^1$ and

$$\int_0^{2\pi} \{|1 + h(e^{i\theta})| + |1 - h(e^{i\theta})| - 2\} |f(e^{i\theta})| \, d\theta = 0.$$

The integrand is nonnegative, since the sum of the distances from $h(e^{i\theta})$ to 1 and (-1) is not less than 2. Since $f(e^{i\theta})$ does not vanish on any set of positive measure, it follows that

$$-1 \le h(e^{i\theta}) \le 1 \qquad \text{a.e.} \tag{13}$$

Thus, in view of the Poisson representation for H^1 functions (Theorem 3.1), $h(z)$ is real everywhere in $|z| < 1$, hence is identically constant. But (12) and the fact that $\|f\| = 1$ then give

$$|1 + h| = |1 - h|,$$

so $h = 0$. This shows $g(z) \equiv 0$, proving that f is an extreme point.

Conversely, let $f = IF$ be an H^1 function with nonconstant inner factor I and $\|f\| = 1$. To show that f is not an extreme point, we shall construct a function $g \ne 0$ in H^1 such that $\|f + g\| = \|f - g\| = 1$. This will be achieved if $h = g/f$ has the properties (13) and

$$\int_0^{2\pi} h(e^{i\theta}) |f(e^{i\theta})| \, d\theta = 0. \tag{14}$$

Let $J(z) = e^{i\alpha} I(z)$ and define

$$g = \tfrac{1}{2} f(J + 1/J). \tag{15}$$

Then $g \in H^1$, and (13) is obviously satisfied. For (14) also to hold, we have only to choose α such that

$$\int_0^{2\pi} \text{Re}\{e^{i\alpha} I(e^{i\theta})\} |f(e^{i\theta})| \, d\theta = 0. \tag{16}$$

This is possible, since the integral is a real-valued continuous function of α which either vanishes identically or changes sign in the interval $0 \le \alpha \le \pi$. Hence the proof is complete.

COROLLARY. (i) If $f \in H^1$, $\|f\| = 1$, and f is not an extreme point of S, then there are two extreme points f_1 and f_2 such that $f = \frac{1}{2}(f_1 + f_2)$.

(ii) If $\|f\| < 1$, then f is a convex combination of two extreme points.

PROOF. (i) If $\|f\| = 1$ and $f = IF$ is not an outer function, define g by (15) and (16). Then $\|f \pm g\| = 1$, so it will be enough to show that $f \pm g$ are outer functions. But we claim that $f \pm \lambda g$ are outer functions whenever $\lambda \geq 1$. Indeed,

$$f \pm \lambda g = \frac{f}{2J}[2J \pm \lambda(J^2 + 1)]$$

$$= \frac{\lambda}{2} e^{-i\alpha} F\left[\pm J^2 + \frac{2}{\lambda} J \pm 1\right]$$

$$= \pm \frac{\lambda}{2} e^{-i\alpha} F(1 \pm e^{i\beta}J)(1 \pm e^{-i\beta}J),$$

where $\lambda \cos \beta = 1$. The last two factors are outer functions, because they have positive real part (see Exercise 1, Chapter 3). Thus $f \pm \lambda g$ is outer, since it is the product of outer functions.

(ii) If $\|f\| < 1$ and f is not an outer function, let g be as above and choose $\lambda_1 > 1$ and $\lambda_2 > 1$ such that

$$\|f + \lambda_1 g\| = \|f - \lambda_2 g\| = 1.$$

This is possible, by continuity, since $\|f\| < 1$. Then $(f + \lambda_1 g)$ and $(f - \lambda_2 g)$ are extreme points, as shown in (i). If f is outer, the functions $\pm f/\|f\|$ are extreme points.

The next theorem identifies the extreme points of the unit sphere in H^∞. By way of motivation, let us first note that $f \in H^\infty$ is an extreme point if $\|f\| = 1$ and $|f(e^{i\theta})| = 1$ on some set E of positive measure. Indeed, if $g \in H^\infty$ and $\|f + g\|_\infty = \|f - g\|_\infty = 1$, then $g(e^{i\theta}) = 0$ a.e. on E, which implies $g = 0$.

THEOREM 7.9. A function $f \in H^\infty$ with $\|f\| = 1$ is an extreme point of the unit sphere in H^∞ if and only if

$$\int_0^{2\pi} \log(1 - |f(e^{i\theta})|) \, d\theta = -\infty. \tag{17}$$

PROOF. Let $g \in H^\infty$ such that

$$\|f + g\| = \|f - g\| = 1.$$

Then

$$|g(z)|^2 \leq 1 - |f(z)|^2 \leq 2(1 - |f(z)|),$$

and it follows from (17) that

$$\int_0^{2\pi} \log|g(e^{i\theta})| \, d\theta = -\infty,$$

which implies $g = 0$. Hence f is an extreme point.

Conversely, if the integral (17) converges, let

$$g(z) = \exp\left\{ \frac{1}{2\pi} \int_0^{2\pi} \frac{e^{it} + z}{e^{it} - z} \log(1 - |f(e^{it})|) \, dt \right\}.$$

Then $|g(z)| \leq 1$ and

$$|g(e^{i\theta})| \leq 1 - |f(e^{i\theta})| \qquad \text{a.e.}$$

Thus $\|f + g\| \leq 1$ and $\|f - g\| \leq 1$, so f is not an extreme point.

EXERCISES

1. Show that for each fixed ζ, $|\zeta| < 1$, and for each positive integer n, $\phi(f) = f^{(n)}(\zeta)$ is a bounded linear functional on H^p, $p < 1$.

2. Prove Corollary 1 to Theorem 7.7.

3. Prove Corollary 2 to Theorem 7.7.

4. Show that every Hilbert space is uniformly convex.

5. Show that in a uniformly convex space, every boundary point of the unit sphere is an extreme point.

6. Show that $f \in L^\infty$ with $\|f\| = 1$ is an extreme point of the unit sphere in L^∞ if and only if $|f(e^{i\theta})| = 1$ a.e.

7. Show that the unit sphere in L^1 has no extreme points.

NOTES

A. E. Taylor [2, 3] was among the first to study H^p ($1 \leq p \leq \infty$) as a Banach space. He represented the linear functionals (Theorem 7.3) in his paper [3]. Beurling proved Theorem 7.4 for H^2 in his fundamental paper [1]. The proof given in the text is essentially his, suitably extended to $1 \leq p < \infty$. Actually, Beurling showed that every subspace of H^2 invariant under multiplication by z has the form $\mathscr{P}[f_0]$ for some (unique) inner function f_0; and he described the lattice structure of these invariant subspaces. Equivalently, he described the invariant subspaces of the shift operator on ℓ^2. Subsequently, a large literature has evolved on invariant subspaces. References up to 1964 may be found in Helson [3]. Gamelin [1] has extended Beurling's theory to H^p with $0 < p < 1$.

Theorem 7.5 has a long history. Day [1] showed that L^p $(p < 1)$ has no continuous linear functionals except the zero functional. Walters [1, 2] obtained a partial representation of the continuous linear functionals on H^p $(p < 1)$, and Romberg [1] completed all but the case $p = (n + 1)^{-1}$. The full theorem and a number of other results on the dual space structure of H^p are in the paper of Duren, Romberg, and Shields [1]. In particular, this paper contains the failure of the Hahn–Banach separation theorem (Theorem 7.7), as well as an example (constructed via gap series) of a subspace of H^p which has the separation property but does not have the Hahn–Banach extension property. For *invariant* subspaces, however, the two properties are equivalent. Some related results also appear in the thesis of J. Shapiro [1] and in a paper of Duren and Shields [1]. Theorem 7.7 shows that H^p $(p < 1)$ is not locally convex, a fact which Livingston [1] and Landsberg [1] observed more directly. Theorem 7.6 and the lemmas preceding it are due to H. S. Shapiro [2]. A singular function μ with modulus of continuity $O(t \log 1/t)$ can be produced as the Lebesgue function over an appropriate Cantor set with variable ratio of dissection. Another construction uses Riesz products; see Duren [1].

Theorems 7.8 and 7.9 are in a paper of deLeeuw and Rudin [1]. They attribute Theorem 7.9 to Arens, Buck, Carleson, Hoffman, and Royden. Forelli [4] and Gamelin and Voichick [1] have extended Theorem 7.8 in an interesting way to H^1 spaces over compact bordered Riemann surfaces. Gamelin and Voichick also extend Theorem 7.9 to this setting. See also Voichick [3]. For the uniform convexity of L^p $(1 < p < \infty)$, see Köthe [1], p. 358 ff. Newman [4] showed that H^1, though it is not uniformly convex, does have a certain weak form of this property.

Another interesting problem is to characterize the linear isometries of H^p onto H^p. deLeeuw, Rudin, and Wermer [1], and independently Nagasawa [1] did so for $p = \infty$ and for $p = 1$. Forelli [2] recently generalized these results to all $p \neq 2$. If $0 < p \leq \infty$ and $p \neq 2$, then every linear isometry T of H^p onto itself has the form

$$T(f)(z) = \alpha[\varphi'(z)]^{1/p}f(\phi(z)),$$

where α is some complex number of modulus one and φ is some conformal mapping of the unit disk onto itself. The Hilbert space H^2 of course has many other isometries. Forelli also describes the isometries of H^p *into* itself for $p \neq 2$. These are more complicated.

We are now prepared to discuss the theory of extremal problems in H^p spaces. It turns out that extremal problems come in pairs, and that even a partial solution to one problem may help solve the other. We first discuss the existence and uniqueness of solutions to the two types of problems, and establish an important "duality relation." Then we find the qualitative form of the solutions in the case of a rational kernel, which is general enough to include most of the interesting examples. These results offer a systematic and practical method for solving various extremal problems. At the end of the chapter, a few examples are discussed to illustrate the technique.

8.1. THE EXTREMAL PROBLEM AND ITS DUAL

The most general bounded linear functional on H^p ($1 \leq p < \infty$) can be expressed in the form

$$\phi(f) = \frac{1}{2\pi i} \int_{|z|=1} f(z)k(z) \, dz, \tag{1}$$

where $k(e^{i\theta}) \in L^q$, $1/p + 1/q = 1$. The functionals of greatest interest are generated by kernels $k(e^{i\theta})$ which are boundary values of rational functions. Some examples are given below.

(i) $\quad k(z) = \sum_{j=1}^{n} \dfrac{c_j}{z - \beta_j}, \quad |\beta_j| < 1; \quad \phi(f) = \sum_{j=1}^{n} c_j f(\beta_j).$

(ii) $\quad k(z) = n!(z - \beta)^{-n-1}, \quad |\beta| < 1; \quad \phi(f) = f^{(n)}(\beta).$

(iii) $\quad k(z) = \sum_{j=0}^{n} c_j z^{-j-1}; \quad \phi(f) = \sum_{j=0}^{n} c_j a_j,$

where $f(z) = \sum_{j=0}^{\infty} a_j z^j$.

For fixed $k \in L^q$, the typical extremal problem is to find

$$\|\phi\| = \sup_{f \in H^p,\ \|f\|_p \le 1} |\phi(f)|. \tag{2}$$

In addition to finding the value of the supremum, several questions are to be considered. Is the supremum attained, and if so, is the extremal function unique? What are the extremal functions?

A function $h \in L^q$ is said to be *equivalent* to the given kernel k (written $h \sim k$) if h and k belong to the same coset of L^q/H^q; that is, if $h - k \in H^q$. Thus h and k determine the same functional on H^p if and only if $h \sim k$. An application of Theorem 7.1 gives at once the important relation

$$\sup_{f \in H^p,\ \|f\|_p \le 1} \frac{1}{2\pi} \left| \int_{|z|=1} f(z)k(z)\, dz \right| = \min_{g \in H^q} \|k - g\|_q \tag{3}$$

for $1 \le p < \infty$. This is called the *duality relation*. It connects the original extremal problem with what is called the *dual extremal problem*: to find the function $g \in H^q$ which is closest to the given kernel $k \in L^q$ (or, equivalently, to find the function $h \sim k$ of minimal norm). According to Theorem 7.1, the minimum is actually attained; that is, the dual extremal problem always has a solution if $1 < q \le \infty$.

The argument breaks down for $p = \infty$ ($q = 1$), since the conjugate space of L^∞ is larger than L^1. However, the dual extremal problem still has a solution in this case, and the duality relation remains true. To prove this, let C be the subspace of L^∞ consisting of all continuous periodic functions, and let P be the subspace of C generated by $1,\ e^{i\theta},\ e^{2i\theta},\ \dots$. The idea is to view ϕ as a functional on C. By Theorem 7.1,

$$\sup_{f \in P,\ \|f\|_\infty \le 1} |\phi(f)| = \min_{\psi \in P^\perp} \|\phi + \psi\|, \tag{4}$$

where $P^{\perp}$ indicates the subspace of C^* which annihilates P. It is easy to describe $P^{\perp}$. According to the Riesz representation theorem, each $\psi \in C^*$ has the form

$$\psi(f) = \frac{1}{2\pi} \int_0^{2\pi} f(e^{i\theta})\, d\mu(\theta)$$

for some function $\mu(\theta)$ of bounded variation. But $\psi \in P^{\perp}$ implies $\psi(e^{in\theta}) = 0$, $n = 0, 1, 2, \ldots$. Hence it follows from the theorem of F. and M. Riesz that

$$d\mu(\theta) = g(e^{i\theta})e^{i\theta}\, d\theta, \qquad g \in H^1,$$

so that

$$\psi(f) = \frac{1}{2\pi i} \int_{|z|=1} f(z)g(z)\, dz, \qquad g \in H^1.$$

Conversely, every ψ of this form annihilates P. Thus the relation (4) takes the form

$$\sup_{f \in P,\, \|f\|_\infty \le 1} |\phi(f)| = \min_{g \in H^1} \|k + g\|_1. \tag{5}$$

In particular, the dual extremal problem has a solution in the case $q = 1$. On the other hand, for every $f \in H^\infty$ with $\|f\| \le 1$ and for every $g \in H^1$, it is trivially true that $|\phi(f)| \le \|k + g\|_1$. Hence

$$\sup_{f \in P,\, \|f\| \le 1} |\phi(f)| \le \sup_{f \in H^\infty,\, \|f\| \le 1} |\phi(f)| \le \min_{g \in H^1} \|k + g\|.$$

In view of (5), then, equality must hold throughout, and the duality relation is true even for $p = \infty$.

Let us now return to the original extremal problem (2) and show that a solution always exists for $1 < p \le \infty$. In other words, for each L^q function $k(e^{i\theta})$, the supremum in (3) is attained. The idea of the proof is to regard

$$\psi(k) = \frac{1}{2\pi i} \int_{|z|=1} f(z)k(z)\, dz \tag{6}$$

as a bounded linear functional on L^q generated by a fixed function $f(e^{i\theta})$ which belongs (more generally) to L^p. Then $\|\psi\| = \|f\|$. For $1 \le q < \infty$, each $\psi \in (L^q)^*$ has the form (6) for some $f \in L^p$; and those ψ which annihilate H^q are precisely the functionals generated by functions $f \in H^p$. This implies that the extremal problem in question has a solution, since

$$\sup_{\psi \in (H^q)^\perp,\, \|\psi\| \le 1} |\psi(k)|$$

is attained for each fixed $k \in L^q$, by Theorem 7.2.

The proof fails in the case $p = 1$, and for a very good reason: the extremal problem (2) *need not have a solution* in H^1. A counterexample will be given in Section 8.3.

A solution does exist in H^1 if the given kernel $k(e^{i\theta})$ is *continuous*. To prove this, consider again the space C of continuous periodic functions and the subspace P which is the uniform closure of the polynomials in $e^{i\theta}$. Each function $f \in L^1$ defines, by (6), a functional $\psi \in C^*$ with $\|\psi\| = \|f\|$. The argument given above, using the F. and M. Riesz theorem, shows that the annihilator of P consists of all $\psi \in C^*$ of the form (6) with $f \in H^1$. Theorem 7.2 may therefore be invoked to conclude that the supremum in (3) is attained, which was to be shown. Observe that periodicity of the kernel is not really necessary; continuity on the interval $[0, 2\pi]$ is sufficient. The conclusion obviously may be extended to any kernel which is the sum of a continuous function and an H^∞ function.

8.2. UNIQUENESS OF SOLUTIONS

It will be convenient to reserve the term *extremal function* to indicate a solution to the original extremal problem (2). A function $K(e^{i\theta}) \sim k(e^{i\theta})$ for which

$$\|K\|_q = \inf_{h \sim k} \|h\|_q = \inf_{g \in H^q} \|k - g\|_q$$

will be called an *extremal kernel*. Having established the existence of extremal functions and extremal kernels, it is natural to ask about uniqueness. Since $|\phi(e^{i\alpha}f)| = |\phi(f)|$ for every real α, some normalization clearly must be imposed before an extremal function can be unique. It is convenient to single out the extremal functions for which $\phi(f) = \|\phi\|$; these will be called *normalized extremal functions*. To avoid trivial complications, it is always assumed that ϕ is not the zero functional; i.e., $k \notin H^q$. The main results on existence and uniqueness can be stated as follows.

THEOREM 8.1 (Main existence and uniqueness theorem). For each p $(1 \le p \le \infty)$ and for each function $k(e^{i\theta}) \in L^q$ $(1/p + 1/q = 1)$ with $k \notin H^q$, the duality relation

$$\sup_{f \in H^p, \|f\|_p \le 1} |\phi(f)| = \inf_{g \in H^q} \|k - g\|_q$$

holds, where $\phi(f)$ is defined by (1). If $p > 1$, there is a unique extremal function f for which $\phi(f) > 0$. If $p = 1$ and $k(e^{i\theta})$ is continuous, at least one extremal function exists. If $p > 1$ $(q < \infty)$, the dual extremal problem has a unique solution. If $p = 1$ $(q = \infty)$, the dual extremal problem has at least one solution; it is unique if an extremal function exists.

The existence of solutions was established in Section 8.1; only the uniqueness assertions remain to be proved. The argument is based upon the following simple observation.

LEMMA. In order that a function $F \in H^p$, with $\|F\| = 1$ and $\phi(F) > 0$, be an extremal function and that K ($K \sim k$) be an extremal kernel, it is necessary and sufficient that

$$e^{i\theta} F(e^{i\theta}) K(e^{i\theta}) \geq 0 \qquad \text{a.e.} \tag{7}$$

and that

$$|K(e^{i\theta})| = \|K\| \qquad \text{a.e.} \qquad\qquad\qquad \text{if } p = 1;$$

$$|F(e^{i\theta})|^p = \|K\|^{-q} |K(e^{i\theta})|^q \quad \text{a.e.} \qquad\qquad \text{if } 1 < p < \infty; \tag{8}$$

$$|F(e^{i\theta})| = 1 \quad \text{a.e.} \quad \text{on } \{\theta : K(e^{i\theta}) \neq 0\} \quad \text{if } p = \infty.$$

PROOF OF LEMMA. It is clear from the duality relation that F is a normalized extremal function and K an extremal kernel if and only if $\phi(F) = \|K\|$. The lemma then merely expresses the conditions for equality in Hölder's inequality. In the condition for $p = 1$, the fact is used that $F(e^{i\theta})$ cannot vanish on a set of positive measure without vanishing identically.

PROOF OF THEOREM. Let F and K be a normalized extremal function and an extremal kernel, respectively. Since $k \in H^q$, $K(e^{i\theta})$ must be different from zero on a set E of positive measure. Through relations (7) and (8), K determines both sgn $F(e^{i\theta})$ and $|F(e^{i\theta})|$ almost everywhere on E, provided $1 < p \leq \infty$. This means that if $1 < p \leq \infty$, all normalized extremal functions coincide on some set of positive measure, hence coincide almost everywhere. In other words, F is unique. For $p = 1$, the argument shows at least that sgn $F(e^{i\theta})$ is determined almost everywhere, if any extremal functions F exist.

To prove the uniqueness of K, let F be a fixed normalized extremal function (uniqueness is not used), K any extremal kernel. According to (7),

$$\text{Re}\{ie^{i\theta} F(e^{i\theta}) K(e^{i\theta})\} = 0 \qquad \text{a.e.}$$

Let $G = k - K$, so that $G \in H^q$; and let $h(z) = izF(z)G(z)$. Then

$$\text{Re}\{h(e^{i\theta})\} = \text{Re}\{ie^{i\theta} F(e^{i\theta}) k(e^{i\theta})\} \qquad \text{a.e.}$$

But since $h \in H^1$ and $h(0) = 0$, the knowledge of $\text{Re}\{h(e^{i\theta})\}$ completely determines $h(z)$, through the Poisson formula. This shows that G, and hence K, is uniquely determined. The proof is valid in the case $p = 1$ ($q = \infty$) *if* an extremal function F exists.

8.3. COUNTEREXAMPLES IN THE CASE $p = 1$

The existence and uniqueness of extremal functions and extremal kernels has now been established for $1 < p \leq \infty$. The situation is entirely different, however, in the case $p = 1$. An extremal function need not exist in H^1, and if it does, it need not be unique up to normalization. At least one extremal kernel must exist, as shown above, but it is not in general unique. Counterexamples will now be given to justify the three negative statements.

(i) To show that an extremal function may fail to exist, consider the example

$$\phi(f) = \frac{1}{2\pi i} \int_{-1}^{1} f(x)\, dx, \qquad f \in H^1.$$

The existence of the integral (in the principal value sense) can be proved by applying Cauchy's theorem with a contour consisting of the segment $-r \leq x \leq r$ and the semicircle $z = re^{i\theta}, 0 \leq \theta \leq \pi$. Since $f \in H^1$, it follows that

$$\lim_{r \to 1} \int_{-r}^{r} f(x)\, dx = -i \int_{0}^{\pi} f(e^{i\theta}) e^{i\theta}\, d\theta,$$

so that $\phi(f)$ has the usual form (1) with a kernel

$$k(e^{i\theta}) = \begin{cases} 0, & -\pi < \theta < 0 \\ -1, & 0 \leq \theta \leq \pi. \end{cases}$$

An equivalent kernel is $K(z) = k(z) + \frac{1}{2}$; hence $\|\phi\| \leq \|K\| = \frac{1}{2}$. This is also a consequence of the Fejér–Riesz inequality (Theorem 3.13). In fact, the example used to prove the sharpness of the Fejér–Riesz inequality, a conformal mapping of $|z| < 1$ onto an elongated rectangle, shows that $\|\phi\| = \frac{1}{2}$. Thus K is an extremal kernel. If there exists an extremal function F, normalized so that $\phi(F) = \frac{1}{2}$, then by (7)

$$e^{i\theta} F(e^{i\theta}) K(e^{i\theta}) \geq 0 \qquad \text{a.e.}$$

Therefore,

$$F(e^{i\theta}) = \begin{cases} e^{-i\theta} |F(e^{i\theta})|, & -\pi < \theta < 0 \\ -e^{-i\theta} |F(e^{i\theta})|, & 0 \leq \theta \leq \pi. \end{cases}$$

Since $F \in H^1$, it follows that

$$0 = \int_{-\pi}^{\pi} e^{i(n+1)\theta} F(e^{i\theta})\, d\theta$$

$$= \int_{-\pi}^{0} e^{in\theta} |F(e^{i\theta})|\, d\theta - \int_{0}^{\pi} e^{in\theta} |F(e^{i\theta})|\, d\theta, \qquad n = 0, 1, 2, \ldots.$$

Consequently,

$$\int_{-\pi}^{0} e^{in\theta}|F(e^{i\theta})|\, d\theta = \int_{0}^{\pi} e^{in\theta}|F(e^{i\theta})|\, d\theta, \qquad n = 0, \pm 1, \pm 2, \ldots.$$

In other words, the two functions

$$\psi_1(\theta) = \begin{cases} |F(e^{i\theta})|, & -\pi \leq \theta \leq 0 \\ 0, & 0 < \theta < \pi, \end{cases}$$

and

$$\psi_2(\theta) = \begin{cases} 0, & -\pi < \theta < 0 \\ |F(e^{i\theta})|, & 0 \leq \theta \leq \pi, \end{cases}$$

have identical Fourier coefficients. This implies $\psi_1(\theta) = \psi_2(\theta)$ a.e., so $F(z) \equiv 0$, which is obviously impossible.

(ii)   The elementary example

$$\phi(f) = \frac{1}{2\pi} \int_{0}^{2\pi} f(e^{i\theta}) e^{-i\theta}\, d\theta, \qquad f \in H^1,$$

shows that a normalized extremal function need not be unique. Here it is obvious that $\|\phi\| = 1$, and that $f(z) = z$ is an extremal function. But a simple calculation shows that for every complex constant α,

$$f(z) = \frac{(z + \alpha)(1 + \bar{\alpha}z)}{1 + |\alpha|^2}$$

is also extremal.

(iii)   A more elaborate example is needed to demonstrate the non-uniqueness of an extremal kernel. In view of the main theorem, this will automatically be another case in which no extremal function exists. Let

$$k(e^{i\theta}) = \begin{cases} 1, & 0 \leq \theta \leq \dfrac{\pi}{2} \\[2mm] -1, & \dfrac{\pi}{2} < \theta \leq \pi \\[2mm] 0, & -\pi < \theta < 0, \end{cases}$$

and let $\phi(f)$ be the functional with kernel k, defined by (1). Obviously, $\|\phi\| \leq 1$. To show that $\|\phi\| = 1$, consider the function $g(z)$ which maps $|z| < 1$ conformally onto the rectangle with vertices $\pm 1 \pm i\varepsilon$, in such a way that $g(1) = -1 - i\varepsilon$, $g(i) = 1$, and $g(-1) = -1 + i\varepsilon$. Let $e^{i\alpha}$ $(0 < \alpha < \pi/2)$ be the point carried into $(1 - i\varepsilon)$. Then $g(e^{i(\pi-\alpha)}) = 1 + i\varepsilon$, by symmetry. It is geometrically clear that the tangent vector

$$ie^{i\theta}g'(e^{i\theta})\begin{cases} \geq 0, & 0 < \theta < \alpha, \\ \leq 0, & \pi - \alpha < \theta < \pi, \end{cases}$$

while $e^{i\theta}g'(e^{i\theta}) \geq 0$ for $\alpha < \theta < \pi - \alpha$. Thus

$$2\pi i\phi(g') = \int_0^{\pi/2} ie^{i\theta}g'(e^{i\theta})\, d\theta - \int_{\pi/2}^{\pi} ie^{i\theta}g'(e^{i\theta})\, d\theta$$

$$= (2 + i\varepsilon) - (i\varepsilon - 2) = 4.$$

On the other hand,

$$\|g'\|_1 = \frac{4(1 + \varepsilon)}{2\pi}.$$

Therefore,

$$\frac{|\phi(g')|}{\|g'\|_1} = \frac{1}{1 + \varepsilon} \to 1 \quad \text{as} \quad \varepsilon \to 0,$$

which implies $\|\phi\| = 1$.

It now follows from the duality theorem that k is itself an extremal kernel, since $\|k\|_\infty = 1$. The object is to produce another extremal kernel; that is, another function $K \sim k$ with $\|K\|_\infty = 1$. Let $K = k + h$, where $h(z)$ maps $|z| < 1$ conformally onto the half-disk $|w| < 1$, $\text{Im}\{w\} > 0$, with $h(1) = -1$, $h(i) = 0$, $h(-1) = 1$. Obviously, $K \sim k$. But

$$|K(e^{i\theta})| = \begin{cases} |h(e^{i\theta})| = 1, & -\pi < \theta < 0; \\ 1 + h(e^{i\theta}) \leq 1, & 0 \leq \theta \leq \pi/2; \\ |-1 + h(e^{i\theta})| \leq 1, & \pi/2 < \theta \leq \pi. \end{cases}$$

Hence $\|K\|_\infty = 1$.

8.4. RATIONAL KERNELS

If the kernel $k(e^{i\theta})$ is the restriction to the unit circle of a rational function $k(z)$, it is possible to describe the extremal functions and extremal kernels more or less explicitly. This is of some interest, because most of the functionals which arise in practice have rational kernels. Some examples were given at the beginning of the chapter.

Let $k(z)$ be analytic in $|z| < 1$ except for poles at $\beta_1, \beta_2, \ldots, \beta_n$ $(|\beta_k| < 1)$, each pole being repeated according to multiplicity. Suppose $k(z)$ is continuous elsewhere in $|z| \leq 1$. Then by Theorem 8.1, a normalized extremal function F

exists for each p, $1 \leq p \leq \infty$. F is unique for $1 < p \leq \infty$, but not necessarily for $p = 1$. An extremal kernel K exists for each p, $1 \leq p \leq \infty$, and is unique (even for $p = 1$, since F exists).

Consider the function $R(z) = zF(z)K(z)$, which is analytic in $|z| < 1$ except perhaps for poles at the points β_k. According to the lemma of Section 8.2, $R(e^{i\theta}) \geq 0$ a.e., because F and K are extremal. Furthermore, since

$$\lim_{r \to 1} \int_a^b |R(re^{i\theta}) - R(e^{i\theta})| \, d\theta = 0$$

for any arc (a, b) on $|z| = 1$, the classical technique of Schwarz reflection may be applied. (See, for instance, Nehari [5], pp. 183–185.) The conclusion is that $R(z)$ has a continuation onto the extended z-plane as a function analytic except perhaps for poles at $\beta_1, \ldots, \beta_n$; $1/\overline{\beta}_1, \ldots, 1/\overline{\beta}_n$. Hence R is a rational function. If $R(\alpha) = 0$ for $|\alpha| < 1$, then $1/\overline{\alpha}$ is a zero of the same multiplicity. Furthermore, any zero on the boundary $|z| = 1$ must be of *even* multiplicity, since $R(z) \geq 0$ there. Thus there are points $\alpha_1, \ldots, \alpha_m$, $|\alpha_i| \leq 1$, in terms of which R has the structure

$$R(z) = Cz \frac{\prod_{i=1}^{m}(z - \alpha_i)(1 - \overline{\alpha}_i z)}{\prod_{i=1}^{n}(z - \beta_i)(1 - \overline{\beta}_i z)}.$$

By the reflection property, the order of the zero (or pole) of R at the origin must be the same as the order of the zero (or pole) at infinity. If μ of the α_i and ν of the β_i are equal to zero, this statement is expressed by the equation

$$\mu - \nu + 1 = (2n - \nu) - (2m - \mu) - 1.$$

Hence $m = n - 1$. It follows, incidentally, that $C > 0$, since $R(z) \geq 0$ on $|z| = 1$.

Some of the α_i which are zeros of F may possibly coincide with numbers β_i. However, K has a pole at each β_i, and in fact its principal part at each pole must coincide with that of k. Let the α_i be renumbered, if necessary, so that $\alpha_1, \ldots, \alpha_s$ are the zeros of K and $\alpha_{s+1}, \ldots, \alpha_\sigma$ those of F in $|z| < 1$. Then $|\alpha_i| = 1$ for $i = \sigma + 1, \ldots, n - 1$. The functions

$$\tilde{K}(z) = K(z) \prod_{i=1}^{s} \frac{1 - \overline{\alpha}_i z}{z - \alpha_i} \prod_{i=1}^{n} \frac{z - \beta_i}{1 - \overline{\beta}_i z}$$

and

$$\tilde{F}(z) = F(z) \prod_{i=s+1}^{\sigma} \frac{1 - \overline{\alpha}_i z}{z - \alpha_i}$$

are analytic and nonvanishing in $|z| < 1$. $\tilde{F} \in H^p$, while $\tilde{K}(z)$ is continuous in $|z| \leq 1$. Furthermore, $\tilde{F}$ and $\tilde{K}$ are outer functions. To prove this, it suffices to show that the product $\tilde{R}(z) = \tilde{F}(z)\tilde{K}(z)$ has no singular factor. In view of the structure of $R(z)$, however, $\tilde{R}(z)$ is a rational function without zeros or poles in $|z| \leq 1$, except perhaps for zeros on $|z| = 1$. Thus the problem reduces to showing that $(z - \alpha)$ is an outer function if $|\alpha| = 1$. But this follows from the fact that $(z - \alpha)^{-1} \in H^p$ for $p < 1$. Consequently,

$$\tilde{F}(z) = e^{i\gamma} \exp\left\{\frac{1}{2\pi} \int_0^{2\pi} \frac{e^{it} + z}{e^{it} - z} \log|F(e^{it})| \, dt\right\}, \tag{9}$$

and similarly for $\tilde{K}(z)$.

The relations (8) between F and K are yet to be used. Suppose first that $1 < p < \infty$. Then

$$|F(z)|^p = \|K\|^{-q}|K(z)|^q \qquad \text{a.e.} \quad \text{on} \quad |z| = 1.$$

But $|F(z)K(z)| = R(z)$ on $|z| = 1$, so

$$|F(z)| = \|K\|^{-1/p}[R(z)]^{1/p}$$

$$= \left(\frac{C}{\|K\|}\right)^{1/p} \prod_{i=1}^{n-1} |1 - \bar{\alpha}_i z|^{2/p} \prod_{i=1}^{n} |1 - \bar{\beta}_i z|^{-2/p}$$

a.e. on $|z| = 1$. Combination of this expression with (9) gives

$$F(z) = A \prod_{i=s+1}^{\sigma} \frac{z - \alpha_i}{1 - \bar{\alpha}_i z} \prod_{i=1}^{n-1} (1 - \bar{\alpha}_i z)^{2/p} \prod_{i=1}^{n} (1 - \bar{\beta}_i z)^{-2/p}, \tag{10}$$

where A is a complex constant. Since $zF(z)K(z) = R(z)$, it follows that

$$K(z) = B \prod_{i=1}^{s} \frac{z - \alpha_i}{1 - \bar{\alpha}_i z} \prod_{i=1}^{n-1} (1 - \bar{\alpha}_i z)^{2/q} \prod_{i=1}^{n} \frac{(1 - \bar{\beta}_i z)^{1-2/q}}{z - \beta_i}. \tag{11}$$

Here the fact has been used that

$$\frac{z - \alpha}{1 - \bar{\alpha}z} \equiv -\alpha \qquad \text{if} \quad |\alpha| = 1. \tag{12}$$

The structural formulas (10) and (11) remain valid in the cases $p = 1$ and $p = \infty$, with the understanding that $1/\infty = 0$. This may be verified by similar calculations based upon the appropriate relations (8). For $p = \infty$, the extremal function takes the simple form

$$F(z) = e^{i\gamma} \prod_{i=s+1}^{\sigma} \frac{z - \alpha_i}{1 - \bar{\alpha}_i z}.$$

The practical value of the formulas (10) and (11) will in general depend upon the actual calculation of the parameters α_i. This can be a very difficult problem.

Suppose, however, that an integer $\sigma \le n - 1$ and numbers B and $\alpha_1, \alpha_2, \ldots,$ α_{n-1} can be found with $|\alpha_i| < 1$ for $1 \le i \le \sigma$ and $|\alpha_i| = 1$ for $\sigma + 1 \le i \le n - 1$, such that for some $s \le \sigma$ the function K given by the formula (11) is equivalent to the original kernel k. Then K is the extremal kernel and, with an appropriate choice of the constant A, F defined by (10) is a normalized extremal function (unique if $p > 1$). This follows from the lemma of Section 8.2.

The problem therefore reduces to an interpolation problem: to find numbers B and α_i, subject to the above restrictions, such that K has the same principal part as k at each of the poles β_i. This problem is guaranteed to have a unique solution if $1 < p \le \infty$, since there is a unique kernel K of the form (11) and a unique normalized extremal function F with the structure (10).

The interpolation problem has at least one solution in the case $p = 1$. Here it is a question of finding numbers B and $\alpha_1, \ldots, \alpha_s$ with $|\alpha_i| < 1$ and $s \le n - 1$, such that

$$K(z) = B \prod_{i=1}^{s} \frac{z - \alpha_i}{1 - \bar{\alpha}_i z} \prod_{i=1}^{n} \frac{1 - \bar{\beta}_i z}{z - \beta_i}$$

has the given principal parts at the poles β_i. This modified extremal problem will have a unique solution, because there is only one extremal kernel. If it turns out that $s = n - 1$, the extremal function F is uniquely determined by (10), with the first product missing. Otherwise (if $0 \le s < n - 1$), the remaining α_i can be chosen *arbitrarily* in the disk $|\alpha| \le 1$. If $|\alpha_i| < 1$ for $s + 1 \le i \le \sigma$ and $|\alpha_i| = 1$ for $\sigma + 1 \le i \le n - 1$, the function (10) (with $p = 1$) is necessarily an extremal function, and every extremal function is obtained in this manner. This characterizes the family of extremal functions in the indeterminate case.

8.5. EXAMPLES

(i) *A coefficient problem and its dual.* Given complex constants $c_0, c_1, \ldots, c_n \, (c_n \neq 0)$, let it be required to find

$$\sup_{f \in H^p, \, \|f\|_p \le 1} |c_0 a_0 + c_1 a_1 + \cdots + c_n a_n|, \tag{13}$$

where $f(z) = \sum_{j=0}^{\infty} a_j z^j$. This coefficient problem was posed and elegantly solved by E. Landau in the special case $p = \infty$ and $c_0 = c_1 = \cdots = c_n = 1$. From the present point of view, it is an extremal problem with the *rational* kernel

$$k(z) = \sum_{j=0}^{n} c_j z^{-j-1}.$$

The dual extremal problem is

$$\inf_{g \in H^q} \|k - g\|_q \qquad \left(\frac{1}{p} + \frac{1}{q} = 1\right),$$

which is equivalent to the problem

$$\inf_{g \in H^q} \left\| \sum_{j=0}^{n} c_j z^{n-j} - z^{n+1} g(z) \right\|_q.$$

In other words, the dual extremal problem is to find $\inf \|h\|$ among all H^q functions of the form

$$h(z) = c_n + c_{n-1} z + \cdots + c_0 z^n + \sum_{j=n+1}^{\infty} a_j z^j.$$

The latter is a "minimal interpolation" problem at the origin. According to the general theory, there is always a unique solution ($1 \le q \le \infty$), and the value of the minimum is equal to the supremum in (13). The problem (13) also has a unique normalized solution if $1 < p \le \infty$, and at least one solution if $p = 1$.

In fact, these solutions take the forms (11) and (10), respectively. The parameters B and α_i must be determined so that $|\alpha_i| \le 1$, $|\alpha_i| < 1$ for $1 \le i \le s$, and

$$B \prod_{i=1}^{s} \frac{z - \alpha_i}{1 - \bar{\alpha}_i z} \prod_{i=1}^{n} (1 - \bar{\alpha}_i z)^{2/q} = c_n + c_{n-1} z + \cdots + c_0 z^n + \cdots. \tag{14}$$

Assuming $p > 1$ and recalling that $c_n \ne 0$, let

$$(c_n + c_{n-1} z + \cdots + c_0 z^n)^{q/2} = \sum_{j=0}^{\infty} \lambda_j z^j$$

for small $|z|$. The numbers $\lambda_0, \lambda_1, \ldots, \lambda_n$ can be expressed in terms of $c_0, c_1, \ldots, c_n$. If

$$P_n(z) = \lambda_0 + \lambda_1 z + \cdots + \lambda_n z^n$$

does not vanish in $|z| < 1$, then

$$P_n(z) = \lambda_0 \prod_{i=1}^{n} (1 - \bar{\alpha}_i z), \qquad |\alpha_i| \le 1,$$

and

$$[P_n(z)]^{2/q} = c_n \prod_{i=1}^{n} (1 - \bar{\alpha}_i z)^{2/q}$$

is the required solution to (14). Thus

$$K(z) = c_n \prod_{i=1}^{n} (1 - \bar{\alpha}_i z)^{2/q} z^{-n-1}$$

is the extremal kernel, and the extremal function has the form

$$F(z) = A \prod_{i=1}^{\sigma} \frac{z - \alpha_i}{1 - \bar{\alpha}_i z} \prod_{i=1}^{n} (1 - \bar{\alpha}_i z)^{2/p},$$

where $\alpha_1, \ldots, \alpha_\sigma$ are the reflections of the zeros of $P_n(z)$, if any, which lie outside the unit circle.

The functions K and F are not so easily found if $P_n(z)$ vanishes in $|z| < 1$. Various criteria are available, however, for concluding that $P_n(z)$ has no zeros inside the unit circle. (See M. Marden [1].) For example, the classical Eneström–Kakeya theorem asserts that this is the case if

$$\lambda_0 \geq \lambda_1 \geq \cdots \geq \lambda_n > 0.$$

If

$$\lambda_0 > \lambda_1 > \cdots > \lambda_n > 0,$$

all the zeros of $P_n(z)$ lie strictly outside the unit circle.

(ii) *Minimal interpolation.* Let distinct points $z_1, \ldots, z_n$ be given in $|z| < 1$. It is required to find an H^p function $(1 \leq p \leq \infty)$ of minimal norm which takes prescribed values

$$f(z_j) = w_j, \qquad j = 1, \ldots, n,$$

at the given points. The problem could be generalized by prescribing the values of the function and its first few derivatives at the points z_j, but the more special problem will serve to illustrate the main ideas. It is not immediately clear that a solution exists at all. However, the problem can be cast in the form of a dual extremal problem with rational kernel, which will ensure the existence and uniqueness of a solution. Let

$$B(z) = \prod_{i=1}^{n} \frac{z - z_i}{1 - \bar{z}_i z} \qquad \text{and} \quad B_j(z) = \prod_{i \neq j} \frac{z - z_i}{1 - \bar{z}_i z}.$$

Define

$$h(z) = \sum_{j=1}^{n} w_j \frac{B_j(z)}{B_j(z_j)}.$$

Then $h(z)$ is analytic in $|z| \leq 1$ and $h(z_j) = w_j$, $j = 1, 2, \ldots, n$. The general H^p function taking the values w_j at z_j is

$$f(z) = h(z) - B(z)g(z),$$

where g is an arbitrary H^p function. The problem of minimizing $\|f\|$ is therefore equivalent to the problem of finding

$$\min_{g \in H^p} \|k - g\|_p,$$

where

$$k(z) = \frac{h(z)}{B(z)} = \sum_{j=1}^{n} \frac{w_j}{B_j(z_j)} \frac{1 - \bar{z}_j z}{z - z_j}.$$

This is a dual extremal problem with the rational kernel k. The associated functional on H^q $(1/p + 1/q = 1)$ is

$$\phi(f) = \frac{1}{2\pi i} \int_{|z|=1} k(z)f(z)\, dz = \sum_{j=1}^{n} \frac{w_j(1 - |z_j|^2)}{B_j(z_j)} f(z_j).$$

The duality relation (3) therefore gives the curious result

$$\min_{f \in H^p,\, f(z_j) = w_j} \|f\|_p = \max_{f \in H^q,\, \|f\|_q \leq 1} \left| \sum_{j=1}^{n} \frac{w_j(1 - |z_j|^2)}{B_j(z_j)} f(z_j) \right|. \qquad (15)$$

(iii) *Extremal problems in H^2*. If the kernel is rational and $p = q = 2$, the extremal kernel (11) takes the form

$$K(z) = B \prod_{i=1}^{s} (z - \alpha_i) \prod_{i=s+1}^{n-1} (1 - \overline{\alpha_i} z) \prod_{i=1}^{n} (z - \beta_i)^{-1}. \qquad (16)$$

The parameters must be chosen so that this function K is equivalent to the given kernel k. But the "natural kernel" k^*, defined as the sum of the principal parts of k at each of the poles β_i, has precisely the form (16):

$$k^*(z) = Q(z) \prod_{i=1}^{n} (z - \beta_i)^{-1},$$

where Q is a polynomial of degree $(n - 1)$ or less. Hence the natural kernel is always the extremal kernel in the H^2 case.

Knowing the extremal kernel, it is easy to find the extremal function F as given by (10), without actually determining the roots α_i. Indeed, in view of (12),

$$z^{n-1}\overline{Q(1/\overline{z})} = C \prod_{i=1}^{s} (1 - \overline{\alpha_i} z) \prod_{i=s+1}^{\sigma} (z - \alpha_i) \prod_{i=\sigma+1}^{n-1} (1 - \overline{\alpha_i} z).$$

Hence

$$F(z) = A z^{n-1}\overline{Q(1/\overline{z})} \prod_{i=1}^{n} (1 - \overline{\beta_i} z)^{-1}.$$

(iv) *The coefficient problem in H^1*. Let us return now to the coefficient problem (13) and its dual, the minimal interpolation problem at the origin. Let m_p be the value of the maximum in (13). Obviously,

$$m_2 = \{|c_0|^2 + |c_1|^2 + \cdots + |c_n|^2\}^{1/2}.$$

It is also possible to find m_1, at least up to an algebraic computation. Any H^1 function $f(z) = \sum a_n z^n$ can be expressed as the product of two H^2 functions g and h with

$$\|f\|_1 = \|g\|_2 \|h\|_2.$$

If

$$g(z) = \sum_{k=0}^{\infty} x_k z^k, \qquad h(z) = \sum_{k=0}^{\infty} y_k z^k,$$

then

$$a_k = x_0 y_k + x_1 y_{k-1} + \cdots + x_k y_0.$$

Thus

$$\sum_{k=0}^{n} c_k a_k = \sum_{\substack{j=0 \\ j+k \le n}}^{n} \sum_{k=0}^{n} c_{j+k} x_j y_k,$$

and the problem reduces to finding

$$m_1 = \max_{\|x\| \le 1, \, \|y\| \le 1} \left| \sum_{j=0}^{n} \sum_{k=0}^{n} A_{jk} x_j y_k \right|,$$

where $\|x\|^2 = |x_0|^2 + \cdots + |x_n|^2$ and (A_{jk}) is the Hankel matrix

$$(A_{jk}) = \begin{bmatrix} c_0 & c_1 & c_2 \cdots c_n \\ c_1 & c_2 & \cdots 0 \\ c_2 & & \\ \vdots & & \\ c_n & 0 & \cdots 0 \end{bmatrix}.$$

The maximum of this symmetric bi-quadratic form can be characterized as the greatest absolute value of the eigenvalues of the matrix (A_{jk}). By the duality relation, this is also the norm of the minimal interpolating function.

EXERCISES

1. Show that if $f(z) = \sum a_k z^k \in H^p$ $(1 < p \le \infty)$, then $|a_n| \le \|f\|_p$, with equality if and only if $f(z) = Az^n$.

2. Show that if $f(z) = \sum a_k z^k \in H^1$, then $|a_n| \le \|f\|_1$, with equality if and only if f has the form

$$f(z) = A \prod_{i=1}^{n} (z - \alpha_i)(1 - \bar{\alpha}_i z),$$

where the α_i are arbitrary complex constants.

3. For distinct points $z_1 = 0$, $z_2, \ldots, z_n$ in $|z| < 1$, and for $0 < p \le \infty$, find the H^p function of minimum norm such that

$$f(0) = 1; \qquad f(z_2) = \cdots = f(z_n) = 0.$$

What is the minimum norm?

4. Show that if $f \in H^p$ $(0 < p \leq \infty)$, then
$$|f(z)| \leq (1 - |z|^2)^{-1/p}\|f\|_p, \qquad |z| < 1.$$
Show that the inequality is sharp for each fixed z.

5. For fixed z, $|z| < 1$, and for each positive integer n, find the maximum of $|f^{(n)}(z)|$ over all H^2 functions f with $\|f\|_2 \leq 1$. Determine the extremal functions.

6. For $f \in H^2$ and for $|z_1| < 1$, $|z_2| < 1$, and $z_1 \neq z_2$, prove
$$\left|\frac{f(z_1) - f(z_2)}{z_1 - z_2}\right| \leq \left\{\frac{1 - |z_1 z_2|^2}{|1 - z_1 \overline{z_2}|^2 (1 - |z_1|^2)(1 - |z_2|^2)}\right\}^{1/2} \|f\|_2,$$
and show that the constant is best possible.

7. For each fixed z, $|z| < 1$, and for each *odd* integer $n = 2m + 1$, prove the sharp inequality
$$|f^{(n)}(z)| \leq \frac{n!}{(1 - |z|^2)^n}\left\{\sum_{k=0}^{m}\binom{m}{k}^2 |z|^{2k}\right\}\|f\|_\infty, \qquad f \in H^\infty,$$
and identify the extremal functions. (*Hint*: Show that the natural kernel is extremal.) (Szász [2]; Macintyre and Rogosinski [2].)

8. Let M_n denote the maximum of $|a_0 + a_1 + \cdots + a_n|$ among all H^∞ functions $f(z) = \sum a_k z^k$ with $\|f\|_\infty \leq 1$. Show that
$$M_n = \sum_{k=0}^{n}(\lambda_k)^2, \qquad \lambda_k = (-1)^k\binom{-\frac{1}{2}}{k} = \frac{(2k)!}{4^k(k!)^2}.$$
Hence $M_n \sim 1/\pi \log n$ as $n \to \infty$. (*Hint*: Observe that
$$\left\{\sum_{k=0}^{n}\lambda_k z^k\right\}^2 = 1 + z + \cdots + z^n + b_{n+1}z^{n+1} + \cdots.)$$
(Landau [1, 2].)

9. Find the distance from H^1 to the function
$$k(z) = (1 - z^{-n-1})(1 - z)^{-1};$$
that is, find the minimum of $\|k - f\|_1$ for all $f \in H^1$.

10. Show that for $f(z) = \sum a_k z^k \in H^1$,
$$\max_{\|f\|_1 \leq 1}|a_0 + a_1 + \cdots + a_n| = \frac{1}{2}\sec\frac{(n+1)\pi}{2n+3}.$$

(*Hint*: Observe that the eigenvalue problem

$$x_{n-k} + x_{n-k+1} + \cdots + x_n = \lambda x_k, \qquad k = 0, 1, \ldots, n,$$

has the solution

$$\lambda = \frac{1}{2} \sec \frac{(n+1)\pi}{2n+3}; \qquad x_k = \sin \frac{(k+1)\pi}{2n+3}, \qquad k = 0, 1, \ldots, n.)$$

(Egerváry [1].)

NOTES

The theory of extremal problems in H^p spaces has evolved from a long history of scattered work on special examples. Landau [1, 2] solved the coefficient problem (i) in Section 8.5 in the case $p = \infty$ and $c_0 = c_1 = \cdots = c_n = 1$. Szász [3, 4] discussed the more general problem in H^∞. Carathéodory and Fejér [2] proposed the "minimal interpolation" problem to minimize $\|f\|_\infty$ among all $f \in H^\infty$ whose Taylor coefficients $a_0, a_1, \ldots, a_n$ at the origin are prescribed. They proved the existence and uniqueness of a solution, and they gave an algebraic method for computing the minimum. Gronwall [1] simplified this work. F. Riesz [1] discussed the same problem in H^1, proved the existence and uniqueness of a solution, and characterized it using a variational method. Fejér [1] solved the coefficient problem (i) of Section 8.5 in H^1, as well as its dual, the Carathéodory–Fejér problem, essentially by the method described in (iv) of Section 8.5. Egerváry [1] obtained a completely explicit solution in the case $c_0 = c_1 = \cdots = c_n = 1$. Kakeya [1, 2] considered the more general interpolation problem (ii) of Section 8.5 and found the form of the solution. In the H^∞ case, this problem was also discussed by Pick [1, 2], Schur [1], and R. Nevanlinna [1]. Geronimus [1, 2, 3] treated some problems similar to those of Carathéodory–Fejér and Riesz. Doob [1] considered the problem of approximating an arbitrary rational kernel by an H^1 function, proved the existence and uniqueness of a solution, and found its qualitative form. Goluzin [1, 2] discussed a wide variety of extremal problems.

The first systematic attempt to unify the theory and to extend it to H^p ($1 \le p \le \infty$) was the paper of Macintyre and Rogosinski [1]. They formulated the duality relation in full generality, proved the existence and uniqueness of extremal functions and extremal kernels, described their form in the case of a rational kernel, and applied the theory to obtain explicit solutions to a great many extremal problems.

Nevertheless, the theory achieved complete unity and elegance only when Havinson [1, 2] and (independently) Rogosinski and Shapiro [1] introduced the methods of functional analysis. A. L. Shields discusses this development in a historical note following his translation of Havinson's paper [2]. Subsequent refinements by H. S. Shapiro [1], Havinson [3], Bonsall [1], and others have led to the main existence and uniqueness theorem as stated and proved in Sections 8.1 and 8.2. The three counterexamples given in Section 8.3 are essentially due to Rogosinski and Shapiro [1]; the third is given in a simplified form suggested by Caughran [1].

Extremal problems for H^p spaces over multiply connected domains (see Chapter 10) have been studied by Ahlfors [1], Garabedian [1], Nehari [2, 3, 4], Rudin [1, 2], Havinson [3], Penez [1], Lax [1], Tumarkin and Havinson [8], and others. Akutowicz and Carleson [1] have studied the analytic continuation properties of extremal interpolatory functions.

Let $z_1, z_2, \ldots$ be distinct points in the open unit disk, and let $w_1, w_2, \ldots$ be arbitrary complex numbers. The general interpolation problem is to characterize the pairs of sequences $\{z_k\}$ and $\{w_k\}$ for which there exists a function $f \in H^p$ with $f(z_k) = w_k$, $k = 1, 2, \ldots$, and to find all of the interpolating functions. But in such general form the problem is difficult, and has not yet been fully solved (see *Notes*). In this chapter we confine attention to the more modest problem of describing the "universal interpolation sequences." These are the sequences $\{z_k\}$ which admit an H^∞ interpolation for every bounded sequence $\{w_k\}$. This restricted problem has a natural generalization to H^p $(0 < p < \infty)$ which we also discuss. The chapter concludes with a theorem of Carleson which generalizes a special result used in proving the main interpolation theorem. Carleson's theorem plays an important part in the proof of the corona theorem (Chapter 12), and it has other applications.

9.1. UNIVERSAL INTERPOLATION SEQUENCES

Consider the problem of finding the most general function $f \in H^p$ such that $f(z_k) = w_k$, $k = 1, 2, \ldots$. If a particular solution f can be found, it is an easy matter to describe the general solution. Indeed, the interpolation f is unique if and only if $\sum (1 - |z_k|) = \infty$, since the difference of two interpolations would vanish at the points z_k. If $\sum (1 - |z_k|) < \infty$, then the general solution is $f + Bg$, where B is the Blaschke product with zeros $\{z_k\}$ and g is an arbitrary H^p function.

In discussing the existence of an H^p interpolation, it is helpful to adopt a more abstract point of view. Fixing p $(0 < p \le \infty)$ and a sequence $\{z_k\}$, consider the linear operator T which assigns the sequence $\{f(z_k)\}$ to each $f \in H^p$. The range of T is clearly the set of all sequences $\{w_k\}$ for which interpolation is possible. The problem is then to characterize the range $T(H^p)$ for each sequence $\{z_k\}$. It is obvious that $T(H^\infty) \subset \ell^\infty$ for every sequence $\{z_k\}$; and it follows from Theorem 5.9 that if $|z_k| \to 1$, then

$$\lim_{k \to \infty} w_k(1 - |z_k|)^{1/p} = 0$$

for each $\{w_k\} \in T(H^p), 0 < p < \infty$.

It is of interest to identify the sequences $\{z_k\}$ for which $T(H^\infty) = \ell^\infty$. These are called *universal interpolation sequences*, since to *every* $\{w_k\} \in \ell^\infty$ there corresponds a function $f \in H^\infty$ with $f(z_k) = w_k$. It is intuitively clear that $\{z_k\}$ cannot be a universal interpolation sequence if the points are "too close together," because it would then be impossible to interpolate a "highly oscillatory" sequence $\{w_k\}$ by a function with "reasonably small" derivative.

This can be made precise. A sequence $\{z_k\}$ is said to be *uniformly separated* if there is a number $\delta > 0$ such that

$$\prod_{\substack{j=1 \\ j \ne k}}^{\infty} \left| \frac{z_k - z_j}{1 - \bar{z}_j z_k} \right| \ge \delta, \qquad k = 1, 2, \ldots .$$

The convergence of the Blaschke product implies, in particular, that $\sum (1 - |z_k|) < \infty$. It turns out that $\{z_k\}$ is a universal interpolation sequence if and only if it is uniformly separated.

This surprisingly sharp result can be extended to an arbitrary H^p space. For this purpose it is convenient to consider a kind of weighted interpolation. Given a sequence $\{z_k\}$, let T_p be the linear operator on H^p $(0 < p \le \infty)$ defined by

$$T_p(f) = \{(1 - |z_k|^2)^{1/p} f(z_k)\}.$$

In particular, $T_\infty = T$. We noted above that if $p < \infty$, T_p maps H^p into the space of sequences which tend to zero, provided only that $\{z_k\}$ has no limit point in the open disk. On the other hand, T_p need not map H^p into ℓ^p, even if $\sum (1 - |z_k|) < \infty$, as we shall see later (Exercises 1 and 2). The following theorem, however, can be proved.

THEOREM 9.1 (Main interpolation theorem). For $0 < p \le \infty$, $T_p(H^p) = \ell^p$ if and only if $\{z_k\}$ is uniformly separated.

9.2. PROOF OF THE MAIN THEOREM

Let us first assume $1 \le p \le \infty$, deferring the case $0 < p < 1$ to Section 9.3. It is obvious that for every sequence $\{z_k\}$, $T = T_\infty$ is a bounded operator from H^∞ to ℓ^∞. More generally, if $\{z_k\}$ is such that $T_p(H^p) \subset \ell^p$, then T_p is a bounded operator from H^p into ℓ^p. This will follow from the closed graph theorem if it can be shown that T_p is closed. But suppose

$$f_n \in H^p, \qquad f_n \to f,$$

and

$$T_p(f_n) = \{(1 - |z_k|^2)^{1/p} f_n(z_k)\} \to w \in \ell^p.$$

Then $f_n(z_k) \to f(z_k)$ for each k, so $T_p(f) = w$. This proves T_p is a closed operator.

Suppose now that $T_p(H^p) = \ell^p$, and let

$$N^p = \{f \in H^p : f(z_k) = 0, k = 1, 2, \ldots\}$$

denote the kernel of T_p. The quotient space H^p/N^p is a Banach space under the usual norm. (See Section 7.1.) Since $T_p(H^p) = \ell^p$, T_p induces a one–one bounded linear operator $\mathscr{T}_p$ from H^p/N^p onto ℓ^p. By the open mapping theorem, $\mathscr{T}_p^{-1}$ is bounded. In other words, corresponding to each $w = \{w_k\} \in \ell^p$, there is a function $f \in H^p$ with

$$(1 - |z_k|^2)^{1/p} f(z_k) = w_k, \qquad k = 1, 2, \ldots, \tag{1}$$

and

$$\|f\| \le M\|w\|,$$

where M is a constant depending only on p. In particular, for each positive integer k there is a function $f_k \in H^p$ with $\|f_k\| \le M$ and

$$f_k(z_j) = \begin{cases} (1 - |z_k|^2)^{-1/p}, & j = k \\ 0, & j \ne k. \end{cases}$$

For $n > k$, let

$$F_{nk}(z) = f_k(z) \prod_{\substack{j=1 \\ j \ne k}}^{n} \frac{1 - \bar{z}_j z}{z - z_j}.$$

Then $F_{nk} \in H^p$ and $\|F_{nk}\| = \|f_k\| \le M$. On the other hand, for any $F \in H^p$, it follows from Theorem 5.9 that

$$|F(z)| \le C\|F\|(1 - |z|^2)^{-1/p},$$

where C depends only on p. Consequently,

$$\left| \prod_{\substack{j=1 \\ j \neq k}}^{n} \frac{1 - \overline{z_j} z_k}{z_k - z_j} \right| = (1 - |z_k|^2)^{1/p} |F_{nk}(z_k)|$$

$$\leq C \|F_{nk}\| \leq CM.$$

This shows that $\{z_k\}$ is uniformly separated if $T_p(H^p) = \ell^p$.

The converse is more difficult. The plan is to prove it first in the special case $p = 2$, then to apply this to establish the general result. Two lemmas will be needed.

LEMMA 1. If $\{z_k\}$ is a uniformly separated sequence, then

$$\sum_{j=1}^{\infty} \frac{(1 - |z_j|^2)(1 - |z_k|^2)}{|1 - \overline{z_j} z_k|^2} \leq A, \qquad k = 1, 2, \ldots,$$

where A is a constant independent of k.

PROOF. A simple calculation gives

$$\left| \frac{z_k - z_j}{1 - \overline{z_j} z_k} \right|^2 = 1 - \frac{(1 - |z_j|^2)(1 - |z_k|^2)}{|1 - \overline{z_j} z_k|^2}$$

$$= 1 - \alpha_{jk}, \qquad \text{say.}$$

Thus

$$\delta^2 \leq \prod_{j \neq k} \left| \frac{z_k - z_j}{1 - \overline{z_j} z_k} \right|^2 = \prod_{j \neq k} (1 - \alpha_{jk})$$

$$\leq \exp\left\{ -\sum_{j \neq k} \alpha_{jk} \right\}.$$

This gives the desired result with $A = 1 - 2 \log \delta$.

LEMMA 2. Let a_{jk} $(j, k = 1, 2, \ldots, n)$ be complex numbers such that $a_{kj} = \overline{a_{jk}}$ and

$$\sum_{j=1}^{n} |a_{jk}| \leq M, \qquad k = 1, 2, \ldots, n.$$

Then for any numbers $x_1, \ldots, x_n$,

$$\left| \sum_{j, k=1}^{n} a_{jk} x_j \overline{x_k} \right| \leq M \sum_{j=1}^{n} |x_j|^2.$$

PROOF. Write

$$|a_{jk} x_j \overline{x_k}| = (|a_{jk}|^{1/2}|x_j|)(|a_{jk}|^{1/2}|x_k|)$$

and apply the Cauchy–Schwarz inequality twice. Since the calculation is straightforward, the details will be omitted.

We return now to the proof of the main interpolation theorem. For f and g in H^2, let

$$(f, g) = \frac{1}{2\pi} \int_0^{2\pi} f(e^{i\theta})\overline{g(e^{i\theta})}\, d\theta.$$

It will be convenient also to use the notation

$$B_n(z) = \prod_{j=1}^{n} \frac{z - z_j}{1 - \overline{z}_j z}$$

$$B_{nk}(z) = \frac{1 - \overline{z}_k z}{z - z_k} B_n(z) = \prod_{\substack{j=1 \\ j \neq k}}^{n} \frac{z - z_j}{1 - \overline{z}_j z}, \qquad k \leq n$$

$$b_{nk} = B_{nk}(z_k).$$

Given $w = \{w_k\} \in \ell^2$ and a uniformly separated sequence $\{z_k\}$, we have to show that a function $f \in H^2$ exists such that $T_2(f) = w$. Let

$$g_{nk}(z) = (1 - |z_k|^2)^{3/2}[B_n(z)]^2(z - z_k)^{-2},$$

and let

$$f_n(z) = \sum_{k=1}^{n} w_k b_{nk}^{-2} g_{nk}(z).$$

Then

$$(1 - |z_k|^2)^{1/2} f_n(z_k) = w_k, \qquad k = 1, 2, \ldots, n,$$

and

$$\|f_n\|^2 = (f_n, f_n) = \sum_{j, k=1}^{n} w_j b_{nj}^{-2} \overline{w_k} \overline{b_{nk}}^{-2}(g_{nj}, g_{nk}). \tag{2}$$

But a calculation based on the residue theorem gives

$$(g_{nj}, g_{nk}) = (1 - |z_j|^2)^{3/2}(1 - |z_k|^2)^{3/2}(1 + z_j \overline{z_k})(1 - z_j \overline{z_k})^{-3}.$$

Since

$$(1 - |z_j|^2)(1 - |z_k|^2) \leq |1 - z_j \overline{z_k}|^2,$$

it follows that

$$|(g_{nj}, g_{nk})| \leq 2(1 - |z_j|^2)(1 - |z_k|^2)|1 - z_j \overline{z}_k|^{-2}.$$

By Lemma 1, this implies

$$\sum_{j=1}^{n} |(g_{nj}, g_{nk})| \leq 2A.$$

Consequently, in view of (2) and Lemma 2,

$$\|f_n\|^2 \leq 2A \, \delta^{-4} \|w\|^2,$$

where the fact that $\{z_k\}$ is uniformly separated ($|b_{nk}| \geq \delta$) has been used. It follows that the functions f_n form a normal family, so that a subsequence tends uniformly in each disk $|z| \leq R < 1$ to a function $f \in H^2$ for which

$$T_2(f) = w \qquad \text{and} \qquad \|f\| \leq C\|w\|, \tag{3}$$

where $C = (2A)^{1/2} \, \delta^{-2}$.

It remains to show that $T_2(f) \in \ell^2$ for every $f \in H^2$. Let $w = \{w_k\}$ be an arbitrary ℓ^2 sequence, and let $h_n(z)$ be the H^2 function of minimal norm such that

$$h_n(z_k) = (1 - |z_k|^2)^{-1/2} w_k, \qquad k = 1, 2, \ldots, n.$$

We have just proved the existence of a function $f \in H^2$ satisfying the conditions (3). Clearly, then,

$$\|h_n\| \leq C\|w\|.$$

Appealing to relation (15) of Chapter 8, we therefore see that each $f \in H^2$ satisfies

$$\left| \sum_{k=1}^{n} \frac{w_k(1 - |z_k|^2)^{1/2}}{b_{nk}} f(z_k) \right| \leq \|h_n\| \, \|f\| \leq C\|w\| \, \|f\|.$$

Taking the supremum over all $w \in \ell^2$ with $\|w\| = 1$, we conclude that

$$\left\{ \sum_{k=1}^{n} (1 - |z_k|^2)|f(z_k)|^2 \right\}^{1/2} \leq C\|f\|,$$

since $|b_{nk}| \leq 1$. In particular, $T_2(f) \in \ell^2$ for every $f \in H^2$.

It is now an easy step to show that $T_p(f) \in \ell^p$ for every $f \in H^p$, $0 < p < \infty$. Simply write

$$f(z) = B(z)[g(z)]^{2/p},$$

where B is a Blaschke product and g is a nonvanishing H^2 function. Then

$$\sum_{k=1}^{\infty} (1 - |z_k|^2)|f(z_k)|^p \leq \sum_{k=1}^{\infty} (1 - |z_k|^2)|g(z_k)|^2$$

$$\leq C^2\|g\|^2 = C^2\|f\|^p.$$

That is,

$$\|T_p(f)\| \le C^{2/p}\|f\|, \qquad f \in H^p. \tag{4}$$

Finally, it has to be shown that $T_p(H^p) \supset \ell^p$ ($1 \le p \le \infty$) if $\{z_k\}$ is uniformly separated. Given $\{w_k\} \in \ell^p$, let $g_n(z)$ be the H^p function of minimal norm which satisfies

$$g_n(z_k) = (1 - |z_k|^2)^{-1/p} w_k, \qquad k = 1, 2, \ldots, n.$$

Appealing again to relation (15) of Chapter 8, we have

$$\|g_n\| = \left| \sum_{k=1}^{n} \frac{w_k(1 - |z_k|^2)^{1/q}}{b_{nk}} f(z_k) \right|$$

for some $f \in H^q$ with $\|f\| = 1$, where q is the conjugate index. Thus, by (4),

$$\|g_n\| \le \delta^{-1} \left\{ \sum_{k=1}^{n} |w_k|^p \right\}^{1/p} \left\{ \sum_{k=1}^{n} (1 - |z_k|^2)|f(z_k)|^q \right\}^{1/q}$$

$$\le \delta^{-1} C^{2/q} \|w\|.$$

Trivial modifications are required in the cases $p = 1$ and $p = \infty$, but the final result is the same. The functions $g_n(z)$ therefore constitute a normal family, and a subsequence converges to a function $g \in H^p$ for which $T_p(g) = w$. Since w was arbitrary, this proves $T_p(H^p) \supset \ell^p$, $1 \le p \le \infty$.

9.3. THE PROOF FOR $p < 1$

We have already shown, even for $p < 1$, that $T_p(H^p) \subset \ell^p$ if $\{z_k\}$ is uniformly separated. On the other hand, the proof that $T_p(H^p) \supset \ell^p$ for $1 \le p \le \infty$ was based on the duality relation, which has no counterpart for $p < 1$. Nevertheless, a direct and surprisingly simple construction is available in this case.

Let $\{z_k\}$ be uniformly separated, and let $w = \{w_k\} \in \ell^p$, $0 < p < 1$. We shall construct a function $f \in H^p$ such that $T_p(f) = w$. This for arbitrary $w \in \ell^p$ will prove $T_p(H^p) \supset \ell^p$. Let

$$b_k(z) = \prod_{\substack{j=1 \\ j \ne k}}^{\infty} \frac{|z_j|}{z_j} \frac{z_j - z}{1 - \bar{z}_j z}$$

and

$$g_k(z) = (1 - \bar{z}_k z)^{-2/p}.$$

Since $|b_k(z_k)| \ge \delta > 0$, the series

$$f(z) = \sum_{k=1}^{\infty} (1 - |z_k|^2)^{1/p} w_k [b_k(z_k)]^{-1} b_k(z) g_k(z)$$

converges uniformly in each disk $|z| \le R < 1$. Hence f is analytic in $|z| < 1$. Furthermore,

$$M_p^p(r, f) \le \delta^{-p} \sum_{k=1}^{\infty} (1 - |z_k|^2) |w_k|^p M_p^p(r, g_k)$$

$$\le \delta^{-p} \sum_{k=1}^{\infty} |w_k|^p < \infty,$$

proving that $f \in H^p$. It is easy to see that $T_p(f) = w$. This shows that $T_p(H^p) = \ell^p$ if $\{z_k\}$ is uniformly separated.

The converse can be proved by virtually the same argument used in the case $1 \le p \le \infty$. Of course, H^p and ℓ^p are no longer Banach spaces if $p < 1$, but they are F-spaces (see Section 6.4) under the translation invariant metrics

$$d(f, g) = \frac{1}{2\pi} \int_0^{2\pi} |f(e^{i\theta}) - g(e^{i\theta})|^p \, d\theta$$

and

$$d(x, y) = \sum_{k=1}^{\infty} |x_k - y_k|^p,$$

respectively. Hence the closed graph and open mapping theorems are still available. It has to be checked that the quotient space H^p/N^p is again an F-space, but this presents no difficulty. In fact, if X is any F-space and S is a closed subspace, X/S is an F-space under the obvious metric.

9.4. UNIFORMLY SEPARATED SEQUENCES

The condition for uniform separation is difficult both to understand intuitively and to verify in practice. In fact, the very existence of uniformly separated sequences is not obvious. It is important to find a more accessible characterization. This is a purely geometric problem which will be discussed independently of the interpolation theory.

LEMMA.

$$\left| \frac{\alpha - \beta}{1 - \bar{\alpha}\beta} \right| \ge \frac{|\alpha| - |\beta|}{1 - |\alpha\beta|} \quad \text{if} \quad |\alpha| < 1, \quad |\beta| < 1.$$

PROOF. It is enough to assume $\alpha = a > 0$. Then

$$\left| \frac{a - re^{i\theta}}{1 - are^{i\theta}} \right|^2 = \frac{a^2 - 2ar\cos\theta + r^2}{1 - 2ar\cos\theta + a^2 r^2} = \frac{A + x}{B + x},$$

where

$$A = \frac{a^2 + r^2}{2ar}; \qquad B = \frac{1 + a^2 r^2}{2ar}; \qquad x = -\cos \theta.$$

The problem is to find the minimum of

$$\varphi(x) = \frac{A + x}{B + x}$$

for $-1 \le x \le 1$. But in this interval

$$\varphi'(x) = \frac{B - A}{(B + x)^2} > 0,$$

since $1 \le A < B$. Hence

$$\varphi(x) \ge \varphi(-1) = \frac{(a - r)^2}{(1 - ar)^2},$$

which proves the lemma.

THEOREM 9.2. If there is a constant $c < 1$ such that

$$1 - |z_{n+1}| \le c(1 - |z_n|), \qquad n = 1, 2, \ldots, \tag{5}$$

then $\{z_n\}$ is uniformly separated. The condition (5) is also necessary if $0 \le z_1 < z_2 < \cdots$.

PROOF. The condition (5) implies

$$1 - |z_j| \le c^{j-k}(1 - |z_k|), \quad j > k; \qquad k = 1, 2, \ldots. \tag{6}$$

In particular, $\sum (1 - |z_j|) < \infty$. It follows from (6) that for $j > k$

$$|z_j| - |z_k| \ge (1 - c^{j-k})(1 - |z_k|)$$

and

$$1 - |z_j z_k| = 1 - |z_j| + |z_j|(1 - |z_k|) \le (1 + c^{j-k})(1 - |z_k|).$$

Hence, by the lemma,

$$\left| \frac{z_k - z_j}{1 - \bar{z}_j z_k} \right| \ge \frac{|z_j| - |z_k|}{1 - |z_j z_k|} \ge \frac{1 - c^{j-k}}{1 + c^{j-k}}, \qquad j > k.$$

For $j < k$ this inequality takes the form

$$\left| \frac{z_k - z_j}{1 - \bar{z}_j z_k} \right| \ge \frac{1 - c^{k-j}}{1 + c^{k-j}}, \qquad j < k.$$

Consequently,

$$\prod_{\substack{j=1 \\ j \neq k}}^{\infty} \left| \frac{z_k - z_j}{1 - \bar{z}_j z_k} \right| \geq \prod_{n=1}^{\infty} \left(\frac{1 - c^n}{1 + c^n} \right)^2 > 0,$$

which shows that $\{z_k\}$ is uniformly separated.

Now suppose $0 \leq z_1 < z_2 < \cdots$ and

$$\prod_{\substack{j=1 \\ j \neq k}}^{\infty} \left| \frac{z_k - z_j}{1 - z_j z_k} \right| \geq \delta > 0, \qquad k = 1, 2, \ldots.$$

Then

$$z_{n+1} - z_n \geq \delta(1 - z_n z_{n+1}), \qquad n = 1, 2, \ldots,$$

so that

$$1 - z_{n+1} \leq 1 - \frac{\delta + z_n}{1 + \delta z_n} \leq (1 - \delta)(1 - z_n).$$

Thus $\{z_n\}$ satisfies (5), as claimed.

Sequences $\{z_n\}$ which satisfy the condition (5) are called *exponential sequences.*

9.5. A THEOREM OF CARLESON

Part of the proof of the main interpolation theorem consisted in showing that if $\{z_n\}$ is uniformly separated and $f \in H^p$ $(0 < p < \infty)$, then

$$\left\{ \sum_{n=1}^{\infty} (1 - |z_n|) |f(z_n)|^p \right\}^{1/p} \leq C \|f\|_p.$$

In other words, if μ is the discrete measure on $|z| < 1$ defined by

$$\mu(z_n) = 1 - |z_n|, \qquad n = 1, 2, \ldots,$$

then

$$\int_{|z| < 1} |f(z)|^p \, d\mu(z) < \infty$$

for each $f \in H^p$, and the injection mapping from H^p to the space L_μ^p (with the obvious definition) is bounded.

It is natural to ask what other measures μ have this property. One trivial example is the ordinary Lebesgue measure on $|z| < 1$. As it turns out, the exact class of admissible measures can be characterized very neatly. A finite measure μ on $|z| < 1$ will be called a *Carleson measure* if there is a constant A such that

$$\mu(S) \leq Ah$$

for every set S of the form

$$S = \{z = re^{i\theta} : 1 - h \leq r < 1; \quad \theta_0 \leq \theta \leq \theta_0 + h\}. \tag{7}$$

The theorem may now be stated as follows.

THEOREM 9.3 (Carleson). Let μ be a finite measure on $|z| < 1$, and suppose $0 < p < \infty$. Then in order that there exist a constant C such that

$$\left\{ \int_{|z| < 1} |f(z)|^p \, d\mu(z) \right\}^{1/p} \leq C \|f\|_p \qquad \text{for all} \quad f \in H^p, \tag{8}$$

it is necessary and sufficient that μ be a Carleson measure.

PROOF OF NECESSITY. It is easy to see that the validity of (8) for any given p is equivalent to its validity for $p = 2$. (Simply factor out Blaschke products and use the standard argument.) Suppose, then, that (8) holds with $p = 2$. Let $z_0 = \rho e^{i\alpha} \neq 0$ be an arbitrary point in the open disk, and consider the H^2 function

$$g(z) = (1 - \overline{z_0} z)^{-1},$$

whose norm is

$$\|g\|_2 = (1 - \rho^2)^{-1/2}.$$

Let $h = 1 - \rho$, and let S be the set of points $z = re^{i\theta}$ such that

$$\rho \leq r < 1 \qquad \text{and} \quad \alpha - h/2 \leq \theta \leq \alpha + h/2.$$

If (8) holds with $p = 2$, then

$$\int_S |g(z)|^2 \, d\mu(z) \leq C^2(1 - \rho^2)^{-1} \leq C^2 h^{-1}. \tag{9}$$

But a simple geometric argument shows that for $z \in S$,

$$|(1/\rho)e^{i\alpha} - z| \leq |(1/\rho)e^{i\alpha} - \rho e^{i(\alpha + h/2)}|$$

$$= \frac{1}{\rho}[1 - 2\rho^2 \cos(h/2) + \rho^4]^{1/2};$$

thus

$$|g(z)|^2 \geq [1 - 2\rho^2 \cos(h/2) + \rho^4]^{-1}$$

$$\geq [1 - 2\rho^2(1 - h^2/4) + \rho^4]^{-1} \geq [5h^2]^{-1}.$$

Combining this with (8), we find

$$\mu(S) \le 5C^2 h,$$

which was to be shown.

PROOF OF SUFFICIENCY. This is much more difficult. We first give an outline of the argument, which is essentially based on the Marcinkiewicz interpolation theorem (see Zygmund [4], Chap. XII), although the special case actually needed is relatively easy, and will be proved independently. The first step is to show that (8) will follow if it can be shown that a certain sublinear operator T is of "type" (2, 2). The second and most difficult step is to show that T is of "weak type" (1, 1). Here the argument relies upon a certain covering lemma, and it is essential that μ be a Carleson measure. The third step, which could be avoided by direct appeal to the Marcinkiewicz theorem, is to deduce that T is of type (2, 2); it is trivially of type (∞, ∞).

First Step. For each $f \in H^2$ we have

$$|f(z)| \le \frac{1}{2\pi} \int_0^{2\pi} P(r, t - \theta)|f(e^{it})| \, dt.$$

Therefore, to show that (8) holds for a given measure μ, it will be sufficient to prove that

$$\left\{ \int_{|z| < 1} [u(z)]^2 \, d\mu(z) \right\}^{1/2} \le C \|\varphi\|_2 \tag{10}$$

if $u(z)$ is the Poisson integral of an L^2 function $\varphi(t) \ge 0$.

With each point $z = re^{i\theta} \ne 0$ in the open disk we associate the boundary arc

$$I_z = \{e^{it} : \theta - \tfrac{1}{2}(1 - r) \le t \le \theta + \tfrac{1}{2}(1 - r)\}.$$

Choose $\theta = 0$ if $z = 0$. (See Figure 3.) It will be convenient to identify the arc I_z with the corresponding segment on the real line (choosing $0 \le \theta < 2\pi$). Given an integrable function $\varphi(t) \ge 0$, periodic with period 2π, define

$$\tilde{\varphi}(z) = \sup_I \frac{1}{|I|} \int_I \varphi(t) \, dt,$$

where the supremum is taken over all intervals I containing I_z, of length $|I| < 1$. It is evident that $\tilde{\varphi}(z)$ is continuous in $0 < |z| < 1$. We shall now prove that

$$u(z) \le 16\pi^2[\tilde{\varphi}(z) + \|\varphi\|_1], \qquad |z| < 1, \tag{11}$$

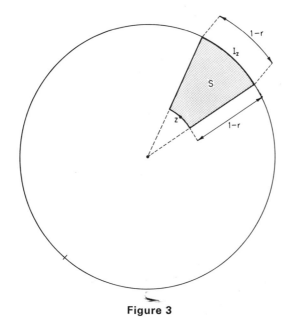

Figure 3

where

$$u(z) = \frac{1}{2\pi} \int_{-\pi}^{\pi} P(r, t - \theta)\varphi(t) \, dt$$

is the Poisson integral of φ. Later, in the third step of the proof, we will show that (10) holds with u replaced by $\tilde{\varphi}$; this together with (11) will prove (10).

To prove (11), fix the point $z = re^{i\theta}$, let $t_n = 2^{n-1}(1 - r)$, and define the intervals

$$\omega_n = [-t_n, t_n], \qquad n = 0, 1, \ldots, N,$$

where N is the largest integer n such that $t_n < \frac{1}{2}$. Now define the sets

$$G_0 = \omega_0 ; \quad G_n = \omega_n - \omega_{n-1}, \qquad n = 1, \ldots, N;$$

$$G_{N+1} = [-\pi, \pi] - \omega_N.$$

Then

$$u(z) = \frac{1}{2\pi} \sum_{n=0}^{N+1} \int_{G_n} P(r, t)\varphi(t + \theta) \, dt$$

$$\leq \frac{1}{2\pi} \sum_{n=0}^{N+1} P(r, t_{n-1}) \int_{G_n} \varphi(t + \theta) \, dt,$$

where $t_{-1} = 0$. But from the inequality

$$P(r, t) \le \frac{1 - r^2}{(1 - r)^2 + 4\pi^{-2}rt^2}, \qquad |t| \le \pi,$$

one finds

$$P(r, t_{n-1}) \le \frac{16\pi^2}{4^n(1 - r)}, \qquad \tfrac{1}{2} \le r < 1.$$

It follows that for $r \ge \tfrac{1}{2}$,

$$u(z) \le \frac{8\pi}{1 - r} \sum_{n=0}^{N+1} 4^{-n} \int_{G_n} \varphi(t + \theta) \, dt$$

$$\le 8\pi \sum_{n=0}^{N} 2^{-n} \tilde{\varphi}(z) + 8\pi^2 \|\varphi\|_1,$$

by the definitions of $\tilde{\varphi}$ and N. If $r < \tfrac{1}{2}$, we have trivially

$$u(z) \le \frac{2}{\pi} \int_{-\pi}^{\pi} \varphi(t + \theta) \, dt = 4\|\varphi\|_1.$$

This proves (11).

Second Step. The operator $T : \varphi \to \tilde{\varphi}$ given by the relation

$$\tilde{\varphi}(z) = \sup_{\substack{I \supset I_z \\ |I| < 1}} \frac{1}{|I|} \int_I |\varphi(t)| \, dt$$

assigns to every periodic $\varphi \in L^1$ a function $\tilde{\varphi}$ continuous in $0 < |z| < 1$. We are going to show that if μ is a Carleson measure, then T is of *weak type* (1, 1):

$$\mu(E_s) \le Cs^{-1}\|\varphi\|_1, \tag{12}$$

where

$$E_s = \{z : \tilde{\varphi}(z) > s\}, \qquad s > 0.$$

For the proof, we define for each $\varepsilon > 0$ the sets

$$A_s^{\varepsilon} = \left\{ z : \int_{I_z} |\varphi(t)| \, dt > s(\varepsilon + |I_z|) \right\}$$

and

$$B_s^{\varepsilon} = \{z : I_z \subset I_w \text{ for some } w \in A_s^{\varepsilon}\}.$$

Observe that the sets B_s^{ε} expand as $\varepsilon \to 0$ and

$$E_s = \bigcup_{\varepsilon > 0} B_s^{\varepsilon}.$$

Hence

$$\mu(E_s) = \lim_{\varepsilon \to 0} \mu(B_s^\varepsilon). \tag{13}$$

If $z_n \in A_s^\varepsilon$ and the arcs I_{z_n} are disjoint, then

$$s \sum_n (\varepsilon + |I_{z_n}|) < \sum_n \int_{I_{z_n}} |\varphi(t)| \, dt \le 2\pi \|\varphi\|_1. \tag{14}$$

In particular, there can be at most a *finite* number of points z_n in A_s^ε whose associated arcs I_{z_n} are disjoint. The following lemma is now needed.

Covering lemma. Let A be a nonempty set in $|z| < 1$ which contains no infinite sequence of points whose associated arcs I_{z_n} are pairwise disjoint. Then there exist a finite number of points $z_1, \ldots, z_m$ in A such that the arcs I_{z_n} are disjoint and

$$A \subset \bigcup_{n=1}^m \{z : I_z \subset J_{z_n}\},$$

where J_z is the arc of length $5|I_z|$ whose center coincides with that of I_z.

Before proving the lemma, let us use it to complete the proof of (12). If E_s is empty, there is nothing to prove. Otherwise, A_s^ε is nonempty for sufficiently small ε, so that

$$A_s^\varepsilon \subset \bigcup_{n=1}^m \{z : I_z \subset J_{z_n}\},$$

where $z_n \in A_s^\varepsilon$ and the arcs I_{z_n} are disjoint. It follows easily that

$$B_s^\varepsilon \subset \bigcup_{n=1}^m \{z : I_z \subset J_{z_n}\}. \tag{15}$$

But since μ is a Carleson measure, we have

$$\mu(\{z : I_z \subset J_{z_n}\}) \le 5A|I_{z_n}|, \qquad n = 1, \ldots, m.$$

Hence, by (15) and (14),

$$\mu(B_s^\varepsilon) \le 5A \sum_{n=1}^m |I_{z_n}| \le 10\pi A s^{-1} \|\varphi\|_1. \tag{16}$$

In view of (13), this completes the proof of (12).

Proof of covering lemma. Let

$$\rho_1 = \inf\{|z| : z \in A\},$$

and choose $z_1 \in A$ with $|z_1| < \frac{1}{2}(1 + \rho_1)$. If $0 \in A$, let $z_1 = 0$. Having chosen $z_1, \ldots, z_{n-1}$, let

$$\rho_n = \inf\{|z| : z \in A \quad \text{and} \quad I_z \cap I_{z_j} = \varnothing, \qquad j = 1, \ldots, n-1\}.$$

Then choose $z_n \in A$ with $I_{z_n} \cap I_{z_j} = \varnothing$ for $j = 1, \dots, n-1$ and $|z_n| < \frac{1}{2}(1 + \rho_n)$. By hypothesis, this process must terminate after the choice of a finite number m of the points z_n. Now for each $z \in A$ we have $I_z \cap I_{z_n} \neq \varnothing$ for some n $(n = 1, \dots, m)$; suppose n is the smallest such index. Then $|z| \geq \rho_n$, which gives

$$1 - |z| \leq 1 - \rho_n < 2(1 - |z_n|).$$

But this together with the fact that I_z meets I_{z_n} implies $I_z \subset J_{z_n}$, which proves the lemma.

Third Step. It now remains to make use of (12) to show that T is of type $(2, 2)$:

$$\int_{|z|<1} |\tilde{\varphi}(z)|^2 \, d\mu(z) \leq K \|\varphi\|_2^2, \qquad \varphi \in L^2. \tag{17}$$

For $s > 0$, let $\varphi = \psi_s + \chi_s$, where

$$\psi_s = \begin{cases} \varphi & \text{wherever } |\varphi| > s/2C; \\ 0 & \text{otherwise.} \end{cases}$$

Here C is the constant in (12); we assume $C \geq 1$. Then

$$\tilde{\varphi}(z) \leq \tilde{\psi}_s(z) + \tilde{\chi}_s(z) \leq \tilde{\psi}_s(z) + \frac{s}{2C},$$

from which it follows that

$$E_s \subset F_s = \left\{ z : \tilde{\psi}_s(z) > \frac{s}{2} \right\}.$$

In particular, $\mu(E_s) \leq \mu(F_s)$. Letting $\alpha(s) = \mu(E_s)$, we therefore have by (12)

$$\int |\tilde{\varphi}|^2 \, d\mu = -\int_0^\infty s^2 \, d\alpha(s) \leq 2 \int_0^\infty s\alpha(s) \, ds$$

$$\leq 2 \int_0^\infty s\mu(F_s) \, ds \leq 4C \int_0^\infty \|\psi_s\|_1 \, ds.$$

But with the notation

$$\sigma_s = \{ t : |\varphi(t)| > s/2C \},$$

we have

$$2\pi \int_0^\infty \|\psi_s\|_1 \, ds = \int_0^\infty \int_{\sigma_s} |\varphi(t)| \, dt \, ds$$

$$= \int_0^{2\pi} |\varphi(t)| \int_0^{2C|\varphi(t)|} ds \, dt = 2C \int_0^{2\pi} |\varphi(t)|^2 \, dt,$$

which proves (17). This completes the proof of the theorem.

REMARK. For later reference (in Chapter 12), we note that if μ is a Carleson measure with constant $A \geq 1$, then the constant C in (8) may be taken to be $4(80)^4 A^2$. This can be verified by a careful examination of the proof.

Finally, it should be observed that the argument, with appropriate modifications, actually proves the following more general theorem.

THEOREM 9.4. Let μ be a finite measure on $|z| < 1$, and suppose $0 < p \leq q < \infty$. Then in order that there exist a constant C such that

$$\left\{ \int_{|z|<1} |f(z)|^q \, d\mu(z) \right\}^{1/q} \leq C\|f\|_p$$

for all $f \in H^p$, it is necessary and sufficient that $\mu(S) \leq Ah^{q/p}$ for every set S of the form (7).

PROOF. The general theorem can be reduced as before to the case $p = 2 \leq q < \infty$. The necessity can then be proved as it was in the case $q = p$.

To prove the sufficiency, one again uses the Marcinkiewicz interpolation theorem (in more general form). What now has to be shown is that under the hypothesis $\mu(S) \leq Ah^q$, the operator $T : \varphi \to \tilde{\varphi}$ is of weak type $(1, q)$ for $q \geq 1$:

$$\mu(E_s) \leq Cs^{-q}\|\varphi\|_1^q, \qquad q \geq 1, \tag{18}$$

for all sets E_s defined as before. (Here we have replaced $q/2$ by q, for notational convenience.) The proof of (18) is the same as for $q = 1$, except that (16) must be replaced by

$$\mu(B_s^\varepsilon) \leq C \sum_{n=1}^m |I_{z_n}|^q \leq C\left\{ \sum_{n=1}^m |I_{z_n}| \right\}^q$$

$$\leq C(2\pi)^q s^{-q}\|\varphi\|_1^q.$$

Theorem 9.4 can be applied to prove the Hardy–Littlewood theorem (Theorem 5.11) that $f \in H^p$ implies

$$\int_0^1 (1 - r)^{q/p-2} M_q^q(r, f) \, dr < \infty, \qquad 0 < p < q < \infty.$$

Another application is a generalization of the Fejér–Riesz theorem (aside from the value of the constant):

$$\left\{ \int_{-1}^1 (1 - r)^{q/p-1}|f(r)|^q \, dr \right\}^{1/q} \leq C\|f\|_p$$

for all $f \in H^p$, $0 < p \leq q < \infty$.

EXERCISES

1. Show that a necessary and sufficient condition for T_p to map H^p *into* ℓ^p $(0 < p \leq \infty)$ is that the measure μ defined by $\mu(z_k) = 1 - |z_k|^2$ be a Carleson measure.

2. Show that if $z_k = 1 - k^{-2}$, then

$$\sum_{k=1}^{\infty} (1 - |z_k|)|f(z_k)|^p = \infty$$

for some $f \in H^p$ $(0 < p < \infty)$, even though $\sum (1 - |z_k|) < \infty$.

3. Show that if $\{z_k\}$ is an exponential sequence, then its associated measure μ (as defined in Exercise 1) is a Carleson measure.

4. Construct a sequence $\{z_k\}$ with $0 < z_1 < z_2 < \cdots$ which is not exponential, but whose associated measure μ is a Carleson measure.

5. Construct a sequence $\{z_k\}$ such that T_p maps H^p into ℓ^p, but not *onto*.

6. Provide the details for the two applications of Theorem 9.4 mentioned at the end of Section 9.5.

7. Use Theorem 9.4 to prove that $f' \in H^{1/2}$ implies $f \in H^1$, a special case of Theorem 5.12.

NOTES

The problem of describing the universal interpolation sequences for H^∞ was proposed by R. C. Buck. After partial solutions by Hayman [2] and by Newman [1], Carleson [1] proved Theorem 9.1 in the case $p = \infty$. Shortly thereafter, Shapiro and Shields [1] gave a simpler proof (which we have essentially followed) and generalized the theorem to $1 \leq p \leq \infty$. The extension to $0 < p < 1$ is due to Kabaïla [2, 5]. Hayman actually constructed an interpolating function under hypotheses a little stronger than uniform separation. P. Beurling later proved the existence of an absolutely convergent interpolation series whenever $\{z_k\}$ is uniformly separated; see Carleson [3].

Theorem 9.2 was found independently by Hayman [2], Kabaïla [1], and Newman [1]. Newman's paper contains a further result of this type. Newman also showed, in an unpublished 1961 manuscript, that $\{z_k\}$ is uniformly separated if and only if the Blaschke product $B(z)$ with zeros z_k has the property that for sufficiently small $\varepsilon > 0$, the level set $|B(z)| = \varepsilon$ decomposes into a system of disjoint curves, each surrounding exactly one point z_k. This justifies the term "uniformly separated". The result recently reappeared in a paper of Hoffman [2].

Nevanlinna and Pick solved the general problem of characterizing the pairs of sequences $\{z_k\}$ and $\{w_k\}$ for which there exists $f \in H^\infty$ with $f(z_k) = w_k$, $k = 1, 2, \dots$. An account of this work is in the paper of R. Nevanlinna [2]. Krein and Rechtman [1] gave another solution based on a certain moment theorem. More recently, Sz.-Nagy and Koranyi [1, 2] based a proof on the spectral theorem for a unitary operator in Hilbert space.

Carleson [1, 2] proved Theorem 9.3 en route to the corona theorem (see Chapter 12). The idea of basing the proof on the Marcinkiewicz interpolation theorem is due to Hörmander [1], although the proof that T is of weak type $(1, 1)$ is similar to Carleson's original argument. The generalization to Theorem 9.4 is in Duren [2]. For a statement and proof of the Marcinkiewicz interpolation theorem, see Zygmund [4].

Up to this point we have dealt almost exclusively with functions analytic in the unit disk. Our objective is now to extend the H^p theory to other domains. Even in the case of a simply connected domain, there are two "natural" generalizations of the space H^p which need not coincide. We begin by presenting these two definitions and exploring the connection between them. Specializing to the case of a Jordan domain with rectifiable boundary, we then discuss boundary behavior and introduce the concept of a Smirnov domain, which we investigate in some detail. We conclude by indicating some extensions of the theory to multiply connected domains. The discussion of H^p spaces over a half plane is reserved for the next chapter.

10.1. SIMPLY CONNECTED DOMAINS

Let D be an arbitrary simply connected domain with at least two boundary points. There is no difficulty in defining the space $H^\infty(D)$ of bounded analytic functions in D; it is a Banach space under the norm

$$\|f\| = \sup_{z \in D} |f(z)|.$$

If $0 < p < \infty$, there are at least two alternatives. One can require the bounded-ness of the integrals of $|f|^p$ over certain curves tending to the boundary, or one can demand that $|f|^p$ have a harmonic majorant (see Section 2.6). The latter definition is simpler, and will be considered first.

A function f analytic in D is said to belong to the class $H^p(D)$ if the sub-harmonic function $|f(z)|^p$ has a harmonic majorant in D. The norm can then be defined as $\|f\| = [u(z_0)]^{1/p}$, where z_0 is some fixed point in D and u is the least harmonic majorant of $|f|^p$. It is easy to see that the space $H^p(D)$ is conformally invariant. That is, if $f \in H^p(D)$ and if $z = \varphi(w)$ is a conformal mapping of a domain D^* onto D, then $f(\varphi(w)) \in H^p(D^*)$. Furthermore, if the norm in $H^p(D^*)$ is defined in terms of the point $w_0 = \varphi^{-1}(z_0)$, this corre-spondence $f \leftrightarrow f \circ \varphi$ is an isometric isomorphism.

A function f analytic in D is said to be of class $E^p(D)$ if there exists a sequence of rectifiable Jordan curves $C_1, C_2, \ldots$ in D, tending to the boundary in the sense that C_n eventually surrounds each compact subdomain of D, such that

$$\int_{C_n} |f(z)|^p \, |dz| \le M < \infty.$$

It is not immediately clear that $E^p(D)$ reduces to the usual H^p space when D is the unit disk. As it turns out, however, it suffices to consider curves C_n which are level curves of an arbitrary conformal mapping of the unit disk onto D.

THEOREM 10.1. Let $\varphi(w)$ map $|w| < 1$ conformally onto D, and let Γ_r be the image under φ of the circle $|w| = r$. Then for each function $f \in E^p(D)$,

$$\sup_{r<1} \int_{\Gamma_r} |f(z)|^p \, |dz| < \infty.$$

PROOF. Let $z_0 = \varphi(0)$, and suppose $\varphi'(0) > 0$. Let $f \in E^p(D)$, and let $\{C_n\}$ be a corresponding sequence of curves. Let D_n denote the interior of C_n. Without loss of generality, suppose z_0 belongs to all of the domains D_n. Let $\varphi_n(w)$ be the conformal mapping of $|w| < 1$ onto D_n, normalized by

$$\varphi_n(0) = z_0, \qquad \varphi_n'(0) > 0.$$

Then $\varphi_n' \in H^1$, and $f(\varphi_n(w))$ is continuous in $|w| \le 1$. By the Carathéodory convergence theorem, φ_n tends to φ uniformly in each disk $|w| \le R < 1$. Thus

$$\int_{\Gamma_r} |f(z)|^p \, |dz| = \int_{|w|=r} |f(\varphi(w))|^p \, |\varphi'(w)| \, |dw|$$

$$= \lim_{n \to \infty} \int_{|w|=r} |f(\varphi_n(w))|^p \, |\varphi_n'(w)| \, |dw|$$

$$\le \liminf_{n \to \infty} \int_{|w|=1} |f(\varphi_n(w))|^p \, |\varphi_n'(w)| \, |dw|$$

$$= \liminf_{n \to \infty} \int_{C_n} |f(z)|^p \, |dz| \le M < \infty.$$

COROLLARY. The following are equivalent:

(i) $f(z) \in E^p(D)$;

(ii) $F(w) = f(\varphi(w)) \, [\varphi'(w)]^{1/p} \in H^p$ for some conformal mapping $\varphi(w)$ of $|w| < 1$ onto D;

(iii) $F(w) \in H^p$ for all such mappings φ.

This last result points out the difference between $H^p(D)$ and $E^p(D)$. In fact, $f \in H^p(D)$ if and only if $f(\varphi(w)) \in H^p$, while $f \in E^p(D)$ if and only if

$$f(\varphi(w))[\varphi'(w)]^{1/p} \in H^p.$$

The two spaces therefore coincide if $|\varphi'(w)|$ is bounded away from 0 and ∞. This will be the case, for example, if D is the interior of an analytic Jordan curve. (See, however, Exercise 3.) It is a surprising fact that the obvious sufficient condition for the equivalence of $H^p(D)$ and $E^p(D)$ is also necessary.

THEOREM 10.2. Let $\varphi(w)$ be an arbitrary conformal mapping of $|w| < 1$ onto the domain D. Then $H^p(D) = E^p(D)$ if and only if there are positive constants a and b such that

$$a \le |\varphi'(w)| \le b, \qquad w \in D. \tag{1}$$

PROOF. It should be observed that (1) is really a property of D, not of φ. We have already shown that (1) implies the equivalence of the two spaces. Conversely, suppose $H^p(D) = E^p(D)$. This means that for functions f analytic in D,

$$f(\varphi(w)) \in H^p \Leftrightarrow f(\varphi(w))[\varphi'(w)]^{1/p} \in H^p.$$

The choice $f(z) \equiv 1$ therefore shows that $\varphi' \in H^1$. On the other hand, if f is chosen so that $f(\varphi(w))[\varphi'(w)]^{1/p} \equiv 1$, it follows that $1/\varphi' \in H^1$. Furthermore, since the modulus of an H^p function can be prescribed arbitrarily on the boundary, subject only to the conditions $f(e^{it}) \in L^p$ and $\log|f(e^{it})| \in L^1$, the hypothesis implies that

$$g \in L^1 \Leftrightarrow \varphi' g \in L^1.$$

(Write $g = g_1 + g_2$, where $\log|g_1| \in L^1$ and $g_2 \in L^\infty$.) But the operator $T: g \to \varphi' g$ from L^1 to L^1 is closed, hence bounded, by the closed graph theorem. In other words, $\varphi'(e^{it}) \in L^\infty$. Since $\varphi' \in H^1$, it follows from Theorem 2.11 that $\varphi' \in H^\infty$, which is half of (1). A similar argument shows that $1/\varphi' \in H^\infty$.

10.2. JORDAN DOMAINS WITH RECTIFIABLE BOUNDARY

Suppose now that D is a domain bounded by a rectifiable Jordan curve C. Let $\varphi(w)$ map $|w| < 1$ conformally onto D, and let $\psi(z)$ be the inverse mapping.

Since C is rectifiable, it makes sense to speak of a tangential direction almost everywhere.

THEOREM 10.3. Each function f of class $E^p(D)$ or $H^p(D)$ has a non-tangential limit almost everywhere on C, which cannot vanish on a set of positive measure unless $f(z) \equiv 0$. Furthermore,

$$\int_C |f(z)|^p \, |dz| < \infty \qquad \text{if} \quad f \in E^p(D),$$

and

$$\int_C |f(z)|^p \, |\psi'(z)| \, |dz| < \infty \qquad \text{if} \quad f \in H^p(D).$$

PROOF. If $f \in E^p(D)$, then

$$F(w) = f(\varphi(w))[\varphi'(w)]^{1/p} \in H^p.$$

As observed in Section 3.5, φ preserves sets of measure zero on the boundary, and it preserves angles at almost every boundary point. But $F(w)$ has a non-tangential limit, and so does $[\varphi'(w)]^{1/p}$, since $\varphi' \in H^1$. Hence $f(\varphi(w))$, and therefore $f(z)$, has a nontangential limit almost everywhere. If $f(z) = 0$ on a boundary set of positive measure, the same is true for $F(w)$; thus $F(w) \equiv 0$, and $f(z) \equiv 0$. The integrability of $|f(z)|^p$ follows from the fact that $|F(w)|^p$ is integrable over $|w| = 1$. The discussion is similar for $f \in H^p(D)$.

One may now ask under what conditions an analytic function f can be recovered from its boundary values by the Cauchy integral formula. In the case of the unit disk, we know (Theorem 3.6) this is true if and only if $f \in H^1$. It will be convenient to use the notation $L^p(C)$ for the class of measurable functions g on C such that $|g(z)|^p$ is integrable with respect to arclength.

THEOREM 10.4. Each $f \in E^1(D)$ has a Cauchy representation

$$f(z) = \frac{1}{2\pi i} \int_C \frac{f(\zeta) \, d\zeta}{\zeta - z}, \qquad z \in D, \tag{2}$$

and the integral vanishes for all z outside C. Conversely, if $g \in L^1(C)$ and

$$\int_C z^n g(z) \, dz = 0, \qquad n = 0, 1, 2, \ldots, \tag{3}$$

then

$$f(z) = \frac{1}{2\pi i} \int_C \frac{g(\zeta) \, d\zeta}{\zeta - z} \in E^1(D), \tag{4}$$

and g coincides almost everywhere on C with the nontangential limit of f.

REMARK. It should be noted that the condition (3) is equivalent to the identical vanishing of the integral (4) outside C, as a series expansion of the Cauchy kernel shows.

PROOF OF THEOREM. If $f \in E^1(D)$, then

$$F(w) = f(\varphi(w))\varphi'(w) \in H^1,$$

so that

$$F(w) = \frac{1}{2\pi i} \int_{|\omega|=1} \frac{F(\omega)\, d\omega}{\omega - w}, \qquad |w| < 1. \tag{5}$$

On the other hand, for fixed w, the function

$$R(\omega) = \frac{\varphi'(w)}{\varphi(\omega) - \varphi(w)} - \frac{1}{\omega - w}$$

is analytic in $|\omega| < 1$ and continuous in $|\omega| \leq 1$. Thus $F(\omega)R(\omega) \in H^1$, and

$$\frac{1}{2\pi i} \int_{|\omega|=1} F(\omega)R(\omega)\, d\omega = 0.$$

Adding this to (5), we obtain

$$F(w) = \frac{\varphi'(w)}{2\pi i} \int_{|\omega|=1} \frac{F(\omega)\, d\omega}{\varphi(\omega) - \varphi(w)},$$

which is equivalent to (2). To prove that the integral vanishes if z is outside C, we need only note that

$$\int_C f(\zeta)\zeta^n\, d\zeta = \int_{|\omega|=1} F(\omega)[\varphi(\omega)]^n\, d\omega = 0, \qquad n = 0, 1, \ldots,$$

since $F \in H^1$.

Conversely, given $g \in L^1(C)$ satisfying (3), the relation (5) allows us to write the function (4) in the form

$$F(w) = \frac{1}{2\pi i} \int_{|\omega|=1} \frac{G(\omega)\, d\omega}{\omega - w} + \frac{1}{2\pi i} \int_C g(\zeta)R(\psi(\zeta))\, d\zeta, \tag{6}$$

where $G(\omega) = g(\varphi(\omega))\varphi'(\omega)$. But $R(\psi(\zeta))$ is analytic in D and continuous in $\bar{D}$, so by Walsh's theorem it can be approximated uniformly in $\bar{D}$ by a polynomial. It now follows from (3) that the second integral in (6) is equal to zero. Hence $F(w)$ is represented as a Cauchy integral which vanishes outside the unit circle (as another application of Walsh's theorem shows). But this implies $F \in H^1$, or $f \in E^1(D)$. The uniqueness of the Cauchy representation (with the integral vanishing identically outside C) is obvious. This completes the proof.

We now recall that in the unit disk the space H^1 coincides not only with the class of Cauchy integrals, but also with the class of analytic Poisson integrals (Theorem 3.1). For a general Jordan domain with rectifiable boundary, the Poisson formula would generalize to *Green's formula*

$$f(z) = \frac{1}{2\pi} \int_C \frac{\partial G(\zeta; z)}{\partial n} f(\zeta) \, |d\zeta|, \tag{7}$$

where $G(\zeta; z)$ is Green's function of D with pole at z and $\partial/\partial n$ indicates the exterior normal derivative. We recall that Green's function is the unique function of the form

$$G(\zeta; z) = \log|\zeta - z| + h(\zeta), \qquad z \in D,$$

where $h(\zeta)$ is harmonic in D and continuous in the closure $\bar{D}$, and $G(\zeta; z) \equiv 0$ for ζ on C. By a *Green integral* over C we mean a function of the form (7), with $f(\zeta)$ replaced by an arbitrary function $k(\zeta)$ for which the integral exists. Just as $E^1(D)$ coincides with the set of Cauchy integrals, it turns out that $H^1(D)$ is the class of analytic Green integrals over the boundary.

THEOREM 10.5. Each $f \in H^1(D)$ satisfies Green's formula (7). Conversely, every Green integral over C which is analytic in D is of class $H^1(D)$.

PROOF. Let $\zeta = \varphi(w)$ map $|w| < 1$ onto D, let $w = \psi(\zeta)$ be its inverse, and let $z = \varphi(0)$. The Green's function of D is then

$$G(\zeta; z) = \log|\psi(\zeta)|,$$

and a straightforward calculation shows that

$$\frac{\partial G(\zeta; z)}{\partial n} = |\psi'(\zeta)|.$$

But $f \in H^1(D)$ implies $f(\varphi(w)) \in H^1$, and it follows that

$$f(z) = \frac{1}{2\pi} \int_{|w|=1} f(\varphi(w)) \, |dw| = \frac{1}{2\pi} \int_C f(\zeta) |\psi'(\zeta)| \, |d\zeta|,$$

which is equivalent to (7). Conversely, if f is the Green integral of k, then

$$\frac{1}{2\pi} \int_C \frac{\partial G(\zeta; z)}{\partial n} |k(\zeta)| \, |d\zeta|$$

will serve as a harmonic majorant of $|f(z)|$.

10.3. SMIRNOV DOMAINS

Again let D be a Jordan domain with rectifiable boundary C, let $z = \varphi(w)$ map D onto $|w| < 1$, and let $w = \psi(z)$ be the inverse mapping. We have seen that every $f \in E^p(D)$ has a boundary function of class $L^p(C)$. By analogy with the result for the unit disk, one might expect this set of boundary functions to coincide with the L^p closure of the polynomials in z. However, this turns out to be true only for a certain subclass of domains which we are about to describe.

Since φ' is an H^1 function with no zeros, it has a canonical factorization of the form

$$\varphi'(w) = S(w)\Phi(w),$$

where S is a singular inner function and Φ is an outer function. D is called a *Smirnov domain* if $S(w) \equiv 1$; that is, if φ' is purely outer. It is easy to check that this is a property only of the domain D, and is independent of the choice of mapping function. Indeed, any other mapping φ_1 has the form $\varphi_1(w) = \varphi(\lambda(w))$, where λ is some linear fractional mapping of the unit disk onto itself. Thus

$$\varphi_1'(w) = \varphi'(\lambda(w))\lambda'(w) = S(\lambda(w))\Phi(\lambda(w))\lambda'(w).$$

But if $S(w)$ is a nontrivial inner function, so is $S(\lambda(w))$. (See Exercise 7, Chapter 2.)

A function $g \in L^p(C)$ will be said to belong to the $L^p(C)$ closure of the polynomials if there is a sequence $\{q_n(z)\}$ of polynomials such that

$$\lim_{n \to \infty} \int_C |g(z) - q_n(z)|^p |dz| = 0.$$

It is convenient to identify $E^p(D)$ with its set of boundary functions. Thus $E^p(D)$ is a closed subspace of $L^p(C)$ which contains all polynomials, hence also their closure.

THEOREM 10.6. Let D be a Jordan domain with rectifiable boundary C, and let $1 \le p < \infty$. Then $E^p(D)$ coincides with the $L^p(C)$ closure of the polynomials if and only if D is a Smirnov domain.

PROOF. If $f \in E^p(D)$, then

$$F(w) = f(\varphi(w))[\varphi'(w)]^{1/p} \in H^p.$$

If D is a Smirnov domain, then $[\varphi']^{1/p}$ is outer, and by Beurling's theorem (Theorem 7.4), there is a polynomial $Q(w)$ such that

$$\|F - Q[\varphi']^{1/p}\| < \varepsilon/2. \tag{8}$$

On the other hand, $Q(\psi(z))$ is analytic in D and continuous in $\bar{D}$, so by Walsh's theorem (see *Notes*) there exists a polynomial $q(z)$ such that

$$|Q(\psi(z)) - q(z)| < (\varepsilon/2)L^{-1/p} \quad \text{on} \quad C,$$

where L is the length of C. This implies

$$\|[Q - q \circ \varphi][\varphi']^{1/p}\| < \varepsilon/2. \tag{9}$$

Combination of (8) and (9) gives

$$\left\{ \int_C |f(z) - q(z)|^p \, |dz| \right\}^{1/p} < \varepsilon,$$

which was to be shown. Conversely, if every $f \in E^p(D)$ can be approximated by polynomials, a similar argument shows that the polynomial multiples of $[\varphi']^{1/p}$ are dense in H^p, so $[\varphi']^{1/p}$ is outer, by Beurling's theorem. Hence φ' is outer, and D is a Smirnov domain. (See Exercise 1.)

If D is not a Smirnov domain, the problem arises to characterize the $L^p(C)$ closure of the polynomials. This can be done in terms of the class N^+ which was discussed in Section 2.5.

THEOREM 10.7. For $1 \leq p < \infty$, a function $f \in E^p(D)$ is in the $L^p(C)$ closure of the polynomials if and only if $f(\varphi(w)) \in N^+$ for some mapping φ of $|w| < 1$ onto D (and therefore for all such mappings).

PROOF. Let $f \in E^p(D)$, so that

$$F(w) = f(\varphi(w))[\varphi'(w)]^{1/p} \in H^p.$$

Thus $f(\varphi(w)) \in N$ in any case, since it is the quotient of two H^p functions. Let $S(w)$ be the singular factor of $\varphi'(w)$. By appeal to Beurling's theorem in its full strength, the proof of Theorem 10.3 may be adapted to show that f belongs to the $L^p(C)$ closure of the polynomials if and only if $[S(w)]^{1/p}$ divides the inner factor of $F(w)$. This is clearly equivalent to saying that $f(\varphi(w)) \in N^+$.

A simple sufficient condition for D to be a Smirnov domain is that

$$\log \varphi'(w) \in H^1.$$

This follows from Corollary 3 to Theorem 3.1. In fact, it is enough that

$$\arg \varphi'(w) \in h^1,$$

since $\log|\varphi'(w)| \in h^1$ whenever D has rectifiable boundary. This follows from the mean value theorem for harmonic functions and the fact that

$$\int_0^{2\pi} \log^+ |\varphi'(re^{i\theta})| \, d\theta \le \int_0^{2\pi} |\varphi'(re^{i\theta})| \, d\theta \le L.$$

In particular, D is a Smirnov domain if arg $\varphi'(w)$ is bounded either from above or from below. Geometrically, this means that the local rotation of the mapping is bounded; loosely speaking, the boundary curve cannot spiral too much. These considerations show that a domain is of Smirnov type if it is starlike (or even "close-to-convex"), or if it has analytic boundary.

Smirnov domains also arise in the study of polynomial expansions of analytic functions. Our next aim is to generalize the simple fact that H^2 of the unit disk is the class of power series with square-summable coefficients.

Associated with the rectifiable Jordan curve C is a unique sequence of polynomials $p_0(z), p_1(z), \ldots$ such that

$$p_n(z) = c_{nn} z^n + c_{n, n-1} z^{n-1} + \cdots + c_{n0}, \qquad c_{nn} > 0,$$

and

$$\frac{1}{L} \int_C p_n(z) \overline{p_m(z)} \, |dz| = \begin{cases} 0, & n \ne m \\ 1, & n = m, \end{cases}$$

where L is the length of C. These are called the *Szegö polynomials* of C; they can be constructed by orthonormalization of the sequence $\{z^n\}$. A full discussion of these orthogonal polynomials and their remarkable connection with conformal mapping is beyond the scope of this book. If C is the unit circle, the nth Szegö polynomial reduces to z^n, $n = 0, 1, \ldots$.

THEOREM 10.8. If D is a Smirnov domain, every function $f \in E^2(D)$ has a unique expansion

$$f(z) = \sum_{n=0}^{\infty} a_n p_n(z), \qquad \text{where} \quad \sum_{n=0}^{\infty} |a_n|^2 < \infty. \tag{10}$$

Furthermore, every series of the form (10) converges uniformly in each closed subdomain of D to a function $f \in E^2(D)$.

PROOF. First let $\{a_n\}$ be an arbitrary square-summable complex sequence. In the space $L^2(C)$, the functions

$$s_n(z) = \sum_{k=0}^{n} a_k p_k(z)$$

form a Cauchy sequence, so there is a function $f \in L^2(C)$ such that

$$\lim_{n \to \infty} \int_C |f(z) - s_n(z)|^2 \, |dz| = 0. \tag{11}$$

Since f is in the $L^2(C)$ closure of the polynomials, it is the boundary function of some $f \in E^2(D)$. By Theorem 10.4, the Cauchy formula

$$f(z) - s_n(z) = \frac{1}{2\pi i} \int_C \frac{[f(\zeta) - s_n(\zeta)]\, d\zeta}{\zeta - z} \qquad (12)$$

is valid. Using the Schwarz inequality, we now see that $s_n(z) \to f(z)$ uniformly in each closed subdomain of D. It should be observed that, for this half of the theorem, D need not be a Smirnov domain. (See Exercise 4, however.)

Conversely, suppose $f \in E^2(D)$, and let

$$a_n = \frac{1}{L} \int_C f(z)\overline{p_n(z)}\, |dz|, \qquad n = 0, 1, \ldots,$$

be the "Fourier coefficients" of f. Among all polynomials of degree n, the "Fourier polynomial" $s_n(z)$ approximates f most closely in the $L^2(C)$ sense. Thus it follows from Theorem 10.6 that (11) holds if D is a Smirnov domain. Applying the Schwarz inequality to the formula (12), we conclude as before that $s_n(z) \to f(z)$ in D. By Parseval's relation,

$$\sum_{n=0}^{\infty} |a_n|^2 = \frac{1}{L} \int_C |f(z)|^2\, |dz|.$$

In particular, this proves the uniqueness of the coefficients.

10.4. DOMAINS NOT OF SMIRNOV TYPE

The notion of a Smirnov domain has been seen to play a decisive role in the theory of approximation and polynomial expansion. The question arises whether there actually exist Jordan domains with rectifiable boundary which are not of Smirnov type. It turns out that non-Smirnov domains do exist, but they are extremely pathological. They can be constructed by an elaborate geometric process due to Keldysh and Lavrentiev. In this section we shall outline a different approach to the problem which shows its close relation to a certain "real-variables" question.

The discussion is based on a remarkably simple criterion for univalence, involving the Schwarzian derivative

$$\{w, z\} = \left(\frac{w''}{w'}\right)' - \frac{1}{2}\left(\frac{w''}{w'}\right)^2.$$

THEOREM (Nehari; Ahlfors–Weill). Let $f(z)$ be analytic in $|z| < 1$, and suppose

$$|\{f(z), z\}| \le k(1 - r^2)^{-2}, \qquad r = |z|, \qquad (13)$$

where $k < 2$. Then $f(z)$ maps $|z| < 1$ conformally onto a Jordan domain (on the Riemann sphere).

The proof is beyond the scope of this book. (See *Notes*.)

For a function $\mu(t)$ of bounded variation over $[0, 2\pi]$, it is convenient to adopt the normalization

$$\mu(t) = \mu(t-), \qquad 0 < t \leq 2\pi.$$

For $t \in [0, 2\pi]$, define

$$v(t) = \mu(t) + Kt, \qquad K = \frac{1}{2\pi}[\mu(0) - \mu(2\pi)], \qquad (14)$$

and let $v(t)$ be extended periodically to $-\infty < t < \infty$. Thus $v(t)$ is continuous on $(-\infty, \infty)$ if and only if $\mu(t)$ is continuous on $[0, 2\pi]$. Let us say that $\mu \in \Lambda_*$ if $v \in \Lambda_*$, in the sense of Chapter 5.

THEOREM 10.9. Let $\mu(t)$ be a normalized real-valued function of bounded variation, and let

$$\mu(t) = \mu_s(t) + \int_0^t w(\tau)\, d\tau$$

be its canonical decomposition into singular and absolutely continuous parts. Let

$$F(z) = \int_0^{2\pi} \frac{e^{it} + z}{e^{it} - z}\, d\mu(t).$$

Then there exists a constant $a > 0$ such that $\exp\{-aF(z)\}$ is the derivative of a function $f(z)$ which maps $|z| < 1$ conformally onto a Jordan domain, if and only if $\mu \in \Lambda_*$. The boundary of this domain is rectifiable if and only if $\mu_s(t)$ is nondecreasing and $\exp\{-2\pi a w(t)\}$ is integrable.

REMARK. This theorem shows, in particular, that the construction of a Jordan domain with rectifiable boundary, the derivative of whose mapping function is a singular inner function alone (as in the Keldysh–Lavrentiev example), is equivalent to the problem of constructing a singular, nondecreasing, bounded function $\mu(t)$ of class Λ_*. This latter construction can be carried out directly. (See *Notes*.)

PROOF OF THEOREM. Set

$$f(z) = \int_0^z \exp\{-aF(\zeta)\}\, d\zeta,$$

so that

$$-\{f(z), z\} = aF''(z) + \frac{a^2}{2}[F'(z)]^2.$$

Let us first observe that there exists a number $a > 0$ such that $f(z)$ maps $|z| < 1$ conformally onto a Jordan domain, if and only if

$$F'(z) = O\left(\frac{1}{1-r}\right). \tag{15}$$

Indeed, if (15) holds, then $F''(z) = O((1-r)^{-2})$, and the inequality (13) can be achieved by a suitably small choice of a. Conversely, if $f(z)$ is univalent in $|z| < 1$, it must satisfy the elementary inequality

$$\left|\frac{zf''(z)}{f'(z)} - \frac{2r^2}{1-r^2}\right| \le \frac{4r}{1-r^2}.$$

(See, e.g., Hayman [1], p. 5 or Nehari [5], p. 216.) This implies (15).

The next step is to show that (15) holds if and only if $\mu \in \Lambda_*$. Let

$$g(z) = \int_0^{2\pi} \frac{e^{it} + z}{e^{it} - z} v(t)\, dt,$$

where v is defined by (14). Integrating by parts, we find

$$F(z) = \int_0^{2\pi} \frac{e^{it} + z}{e^{it} - z} \, dv(t) - 2\pi K = izg'(z) - 2\pi K.$$

Thus

$$F'(z) = ig'(z) + izg''(z).$$

This shows that (15) is equivalent to the condition

$$g''(z) = O\left(\frac{1}{1-r}\right). \tag{16}$$

But according to Theorem 5.3, (16) holds if and only if

$$g \in A \quad \text{and} \quad g(e^{i\theta}) \in \Lambda_*. \tag{17}$$

We claim that g has these properties if and only if $v \in \Lambda_*$. Indeed, suppose $g(z) = u(z) + iv(z)$ satisfies (17). Since $u(z)$ is the Poisson integral of v, $u(e^{i\theta}) = v(\theta)$ wherever v is continuous. Hence, by the normalization of μ, v is continuous everywhere, and is of class Λ_*. Conversely, if $v \in \Lambda_*$, then $u(z)$ is continuous in $|z| \le 1$ and $u(e^{i\theta}) \in \Lambda_*$. Since Λ_* is preserved under conjugation (Theorem 5.8), $v(z)$ has the same properties, so $g(z)$ satisfies (17).

It remains to discuss the rectifiability of the boundary. Here we suppose $f(z)$ maps $|z| < 1$ conformally onto a Jordan domain and has a derivative of the form

$$f'(z) = \exp\left\{ -a \int_0^{2\pi} \frac{e^{it} + z}{e^{it} - z} \, d\mu_s(t) \right\}$$

$$\times \exp\left\{ \frac{1}{2\pi} \int_0^{2\pi} \frac{e^{it} + z}{e^{it} - z} \, [-2\pi a w(t)] \, dt \right\},$$

where $a > 0$, $\mu_s(t)$ is a singular function of bounded variation, and $w(t)$ is integrable. The canonical factorization theorem (Theorem 2.9) makes it clear that $f' \in N$ and that

$$\log|f'(e^{it})| = -2\pi a w(t) \qquad \text{a.e.}$$

It now follows from the factorization theorem for H^1 functions (Theorem 2.8) that $f' \in H^1$ if and only if μ_s is nondecreasing and

$$|f'(e^{it})| = \exp\{-2\pi a w(t)\} \in L^1.$$

Since $f' \in H^1$ is equivalent to the rectifiability of the boundary (Theorem 3.12), the proof is complete.

10.5. MULTIPLY CONNECTED DOMAINS

The definitions of the classes $H^p(D)$ and $E^p(D)$, given in Section 10.1 for a simply connected domain, are easily extended to the case in which D is multiply connected. Suppose, then, that D is an arbitrary domain in the complex plane. For $0 < p < \infty$, let $H^p(D)$ again be the space of analytic functions f such that $|f(z)|^p$ has a harmonic majorant in D. Fixing an arbitrary point $z_0 \in D$, let

$$\|f\| = [u(z_0)]^{1/p}, \qquad f \in H^p(D), \tag{18}$$

where u is the least harmonic majorant of $|f|^p$. As we shall see shortly, this is a genuine norm if $p \geq 1$. It is not difficult to show that different points of reference z_0 induce equivalent norms (see Exercise 2). If D is simply connected (and has at least two boundary points), we know that $f \in H^p(D)$ if and only if $f(z)[\psi'(z)]^{1/p} \in E^p(D)$, where ψ is any conformal mapping of D onto the unit disk. The following theorem may be viewed as a generalization of this result to multiply connected domains.

THEOREM 10.10. Fix a point $z_0 \in D$, let $f(z)$ be analytic in D, and let Δ be any subdomain of D, containing z_0, whose boundary Γ consists of a finite number of continuously differentiable Jordan curves in D. Then $f \in H^p(D)$ if and only if there exists a constant M, independent of Δ, such that

$$\frac{1}{2\pi} \int_\Gamma |f(\zeta)|^p \frac{\partial G(\zeta; z_0)}{\partial n} |d\zeta| \le M, \tag{19}$$

where $G(\zeta; z_0)$ is the Green's function of Δ with pole at z_0.

PROOF. If $|f|^p$ has a harmonic majorant u in D, the integral (19) is always less than or equal to $u(z_0)$. Conversely, suppose (19) holds and let $\{\Delta_k\}$ be an expanding sequence of domains of the type described in the theorem, whose union is D. Let Γ_k denote the boundary of Δ_k, and let $G_k(\zeta; z)$ be the Green's function with pole at $z \in \Delta_k$. Then the function

$$v_k(z) = \frac{1}{2\pi} \int_{\Gamma_k} |f(\zeta)|^p \frac{\partial G_k(\zeta; z)}{\partial n} |d\zeta| \tag{20}$$

is harmonic in Δ_k. Since $|f|^p$ is subharmonic, $v_k(z) \le v_{k+1}(z)$ for all $z \in \Delta_k$; while (19) implies that $v_k(z_0) \le M$ for all k. By Harnack's principle (see Ahlfors [2], p. 236), the sequence $\{v_k\}$ therefore converges to a function u harmonic in D which is clearly a majorant (in fact, the least harmonic majorant) of $|f|^p$.

As a corollary to the proof, we note that if $f \in H^p(D)$, then

$$\|f\| = \lim_{k \to \infty} [v_k(z_0)]^{1/p}, \tag{21}$$

where v_k is defined by (20) and $\|f\|$ by (18). In particular, the limit is independent of the sequence $\{\Delta_k\}$. From (21) and Minkowski's inequality, it follows that $\| \ \|$ is a genuine norm if $p \ge 1$.

The space $H^p(D)$ can be identified in a natural way with a certain subspace of H^p over the unit disk. To develop this correspondence, we first discuss a basic result which is essentially the uniformization theorem for planar domains. (See Goluzin [3], Chap. VI, Section 1; and Ahlfors and Sario [1].)

If D has at least three boundary points, there exists a function $\varphi(w)$ analytic and locally univalent in $|w| < 1$, whose range is precisely D and which is invariant under a group $\mathscr{G}$ of linear fractional mappings of the unit disk onto itself:

$$\varphi(g(w)) \equiv \varphi(w), \qquad g \in \mathscr{G}.$$

Furthermore, if z_0 is an arbitrary point in D, φ may be chosen so that $\varphi(0) = z_0$ and $\varphi'(0) > 0$; and these conditions determine φ uniquely. In other words, the pair $(|w| < 1, \varphi)$ is the *universal covering surface* of D, and $\mathscr{G}$ is the group of *cover transformations*, or the *automorphic group* of D. If $\varphi(w_1) = \varphi(w_2)$, there is some $g \in \mathscr{G}$ such that $g(w_1) = w_2$.

THEOREM 10.11. The mapping

$$f(z) \to F(w) = f(\varphi(w))$$

is an isometric isomorphism of $H^p(D)$ onto the subspace of H^p invariant under $\mathscr{G}$.

REMARK. If D is simply connected, then $\mathscr{G}$ is trivial, and the theorem reduces to an observation already made in Section 10.1.

PROOF OF THEOREM. The function F associated with a given f is clearly invariant under $\mathscr{G}$. If $|f(z)|^p$ has a harmonic majorant $u(z)$, then $u(\varphi(w))$ is a harmonic majorant of $F(w)$. Conversely, let F be an arbitrary H^p function invariant under $\mathscr{G}$, and let $U(w)$ be the *least* harmonic majorant of $|F(w)|^p$. In fact, U is simply the Poisson integral of $|F|^p$ over the unit circle, so $\|F\| = [U(0)]^{1/p}$. Since F is invariant under $\mathscr{G}$, $U(g(w))$ is also the least harmonic majorant of $|F(w)|^p$, and we have $U(g(w)) \equiv U(w)$ for all $g \in \mathscr{G}$. In other words, U is invariant under $\mathscr{G}$. This means that if $\varphi(w_1) = \varphi(w_2)$, then $U(w_1) = U(w_2)$, since there is some $g \in \mathscr{G}$ for which $g(w_1) = w_2$. Hence the function $u(z) = U(\varphi^{-1}(z))$ is well defined and is the least harmonic majorant of $|f(z)|^p$, where $f(z) = F(\varphi^{-1}(z))$. In particular,

$$\|f\| = [u(z_0)]^{1/p} = [U(0)]^{1/p} = \|F\|,$$

so the correspondence between f and F is an isometry.

If D is *finitely* connected, certain questions about functions in $H^p(D)$ can be reduced to the simply connected case by means of a decomposition theorem. For simplicity, we assume that no component of the complement of D reduces to a point, so that (by successive applications of the Riemann mapping theorem) D is conformally equivalent to a domain bounded by a finite number of analytic Jordan curves.

THEOREM 10.12. Let D be a finitely connected domain whose boundary consists of disjoint Jordan curves $C_1, C_2, \ldots, C_n$. Let D_k be the domain with boundary C_k which contains D ($k = 1, 2, \ldots, n$). Then every $f \in H^p(D)$ can be represented in the form

$$f(z) = f_1(z) + f_2(z) + \cdots + f_n(z),$$

where $f_k \in H^p(D_k)$, $k = 1, 2, \ldots, n$.

PROOF. Since $H^p(D)$ is conformally invariant, we may suppose that C_1 is a circle which surrounds $C_2, \ldots, C_n$. Let $\Gamma_1, \Gamma_2, \ldots, \Gamma_n$ be disjoint rectifiable Jordan curves in D which are homologous to $C_1, C_2, \ldots, C_n$, respectively.

Then $f = f_1 + f_2 + \cdots + f_n$, where

$$f_k(z) = \frac{1}{2\pi i} \int_{\Gamma_k} \frac{f(\zeta)}{\zeta - z} \, d\zeta$$

is analytic in D_k (after the obvious analytic continuation). To show that $f_k \in H^p(D_k)$, we need only deal with the case $k = 1$, since any of the curves C_k can be made to play the role of C_1 by a suitable conformal mapping. Let $R \subset D$ be an annulus with outer boundary C_1 and inner boundary Γ, a circle in D. By hypothesis, $|f|^p$ has a harmonic majorant u in D. Thus, since $f_2, \ldots, f_n$ are bounded in R, there is a constant a such that

$$|f_1(z)|^p \le u(z) + a, \qquad z \in R.$$

But $u = u_1 + u_2$, where u_1 is harmonic inside C_1 and u_2 is harmonic outside and on Γ. In particular, u_2 is bounded in R, so

$$|f_1(z)|^p \le u_1(z) + b, \qquad z \in R, \tag{22}$$

for some constant b. But $(u_1 + b)$ is harmonic and $|f_1|^p$ is subharmonic throughout D_1, so the inequality (22) holds throughout D_1. Hence $f_1 \in H^p(D_1)$, and the proof is complete.

If the boundary curves $C_1, C_2, \ldots, C_n$ are rectifiable, it follows at once from the decomposition theorem that a function $f \in H^p(D)$ has a non-tangential limit at almost every boundary point, and the boundary function cannot vanish on a set of positive measure unless $f = 0$.

The classes $E^p(D)$ can also be considered in multiply connected domains. A function f analytic in D is said to belong to $E^p(D)$ if there is a sequence $\{\Delta_\nu\}$ of domains whose boundaries $\{\Gamma_\nu\}$ consist of a finite number of rectifiable Jordan curves, such that Δ_ν eventually contains each compact subset of D, the lengths of the Γ_ν are bounded, and

$$\limsup_{\nu \to \infty} \int_{\Gamma_\nu} |f(z)|^p \, |dz| < \infty.$$

Suppose now that D is finitely connected, and that its boundary C consists of rectifiable Jordan curves $C_1, C_2, \ldots, C_n$. If $n = 1$, we know by Theorem 10.1 that the boundedness of the lengths of the Γ_ν is a superfluous requirement in the definition of $E^p(D)$; but whether this condition can be dispensed with in the multiply connected case is still unknown. For domains D of this type, it is easy to prove a decomposition theorem analogous to Theorem 10.12 for functions $f \in E^p(D)$. It then follows that every such f has a nontangential limit almost everywhere on C, and that (if $p \ge 1$) f can be recovered from its boundary function by a Cauchy integral over C. Also, $E^p(D) = H^p(D)$ if all the boundary curves are analytic.

If D is a finitely connected domain, none of whose boundary components consists of a single point, there is a characterization of $E^p(D)$ which generalizes a result proved earlier for simply connected domains. Let $w = \psi(z)$ be any conformal mapping of D onto a domain bounded by *analytic* Jordan curves, and let $z = \varphi(w)$ be the inverse mapping. If $f(z)$ is analytic in D, then $f \in E^p(D)$ if and only if $|f(\varphi(w))|^p |\varphi'(w)|$ has a harmonic majorant in $G = \psi(D)$. In particular, if $\arg\{\varphi'(w)\}$ is single-valued in G, one can say that $f \in E^p(D)$ if and only if $f(\varphi(w))[\varphi'(w)]^{1/p} \in H^p(G)$. The proof is omitted, since it is similar to the proof of Theorem 10.10.

EXERCISES

1. Let D be a Jordan domain with rectifiable boundary and let φ map $|w| < 1$ conformally onto D. For $0 < p < \infty$, show that every $F \in H^p$ has the form

$$F(w) = f(\varphi(w))[\varphi'(w)]^{1/p}$$

for some $f \in E^p(D)$.

2. Let D be an arbitrary domain, let z_1 and z_2 be any points in D, and let

$$\|f\|_1 = [u(z_1)]^{1/p}, \qquad \|f\|_2 = [u(z_2)]^{1/p}$$

be the corresponding norms on $H^p(D)$. Prove the existence of positive constants A and B such that

$$A\|f\|_1 \leq \|f\|_2 \leq B\|f\|_1, \qquad f \in H^p(D).$$

(*Hint*: Choose a simply connected subdomain containing z_1 and z_2, map it onto the unit disk, and apply the Poisson formula.)

3. Let D be the interior of a Jordan curve which is analytic except at one point, where it has a corner with interior angle α. For $0 < p < \infty$, show that $E^p(D) \subsetneq H^p(D)$ if $0 < \alpha < \pi$; while $H^p(D) \subsetneq E^p(D)$ if $\pi < \alpha < 2\pi$.

4. Let D be a Jordan domain with rectifiable boundary C, not a Smirnov domain. Show that $f \in E^2(D)$ has an expansion

$$f(z) = \sum a_n p_n(z), \qquad \sum |a_n|^2 < \infty,$$

in terms of the Szegö polynomials p_n, if and only if f can be approximated in the $L^2(C)$ sense by polynomials. Hence show that the first statement in Theorem 10.8 is false if D is not a Smirnov domain.

5. Let D be a Jordan domain with rectifiable boundary C, and let $w = \psi(z)$ map D onto $|w| < 1$. Show that $\psi' \in E^1(D)$. Show, however, that $\psi' \notin E^p(D)$ for any $p > 1$ if D is not a Smirnov domain.

6. Let D be a Jordan domain with rectifiable boundary C. Show that if $f \in E^p(D)$ and its boundary function is in $L^q(C)$ for some $q > p$, it need not follow that $f \in E^q(D)$. Show, however, that the conclusion is true if D is a Smirnov domain. (*Hint*: See Exercise 5 and Theorem 2.11.)

7. Let D be a Jordan domain with rectifiable boundary C. According to Theorem 10.4, the set of boundary functions of $f \in E^1(D)$ is precisely the class of functions $f \in L^1(C)$ such that

$$\int_C z^n f(z)\, dz = 0, \qquad n = 0, 1, \ldots .$$

Show that the analogous statement for $1 < p \le \infty$ is true if D is a Smirnov domain, but need not be true otherwise.

8. Show that if φ maps $|w| < 1$ onto a Jordan domain D with rectifiable boundary and $\log|\varphi'(e^{it})| \in L \log^+ L$, then D is a Smirnov domain. Thus D is a Smirnov domain if $|\varphi'|$ is not "too small" on the boundary.

NOTES

For a Jordan domain with rectifiable boundary, the spaces $E^p(D)$ were first considered by Smirnov [3], who defined them in terms of level curves. The equivalence with the apparently more general definition (Theorem 10.1) was proved by Keldysh and Lavrentiev [1]; see Privalov [4]. Smirnov also introduced the definition of $H^p(D)$ and proved Theorems 10.4 and 10.5. The recent book of Smirnov and Lebedev [1] discusses these matters. Theorem 10.2 is due to Tumarkin and Havinson [3]. Privalov [4] gives further information on integrals of Cauchy–Stieltjes type and generalizations of the F. and M. Riesz theorem. A discussion of the Carathéodory convergence theorem and Walsh's theorem can be found in Goluzin [3]. Smirnov gave a "Hilbert space" proof of Theorem 10.6 in the case $p = 2$ (see also Goluzin [3]). Since our proof is based on Beurling's theorem, it is valid even for $p < 1$ (see Gamelin [1]). The general result (for $0 < p < \infty$) was stated by Keldysh [1]. Theorem 10.7 is due to Tumarkin [2], with a different proof; it is also valid for $p < 1$. Theorem 10.8 goes back to Smirnov [1, 3]. For further information on the Szegö polynomials, the reader is referred to Szegö [2, 3].

Smirnov apparently tried without success to prove that every Jordan domain with rectifiable boundary is a Smirnov domain (see Smirnov [3], p. 353), but several years later Keldysh and Lavrentiev [1] produced a counterexample. Their construction was simplified in the book of Privalov [4], but the technical details remain formidable. The relatively simple approach described in Section 10.4, is due to Duren, Shapiro, and Shields [1]. Nehari [1] obtained a Schwarzian derivative criterion for univalence in the open disk, the condition (13) with k replaced by 2. Ahlfors and Weill [1] sharpened it to the form given. See also Duren [5]. Constructions of a singular nondecreasing function of class Λ_* have been carried out by Piranian [1], Kahane [1], and H. S. Shapiro [4]. For sufficient conditions that a domain be of Smirnov type, see Privalov [4], Tumarkin [3], and H. S. Shapiro [3].

Rudin [1, 2] developed the theory of $H^p(D)$ in the multiply connected case and proved Theorems 10.10, 10.11, and 10.12. For the decomposition theorem in $E^p(D)$, see Tumarkin and Havinson [4]. They also found [2] the characterization of $E^p(D)$ mentioned at the end of Section 10.5. Their survey paper [7] gives a clear account of the theory and contains further references.

For a discussion of H^p spaces over an annulus, see Kas'yanyuk [1, 2], Sarason [1], and Coifman and Weiss [1].

There is a large and rapidly growing literature on H^p spaces over Riemann surfaces. Some of the relevant papers are Parreau [1], Royden [1], Voichick [1, 2, 3], Voichick and Zalcman [1], Gamelin and Voichick [1], Forelli [3, 4], Fisher [1], Heins [1], and Earle and Marden [1, 2]. The notes of Heins [2] survey certain aspects of this theory.

Rudin [9] describes some generalizations of the H^p theory to several complex variables, where the domain is a polydisk. Some of the one-variable theory extends to higher dimensions, but there are many counterexamples.

Recently, various aspects of the H^p theory have been generalized to the abstract setting of function algebras. See, for example, the surveys of Gamelin and Lumer [1], Lumer [1], and Gamelin [2], which give further references.

This chapter deals with functions analytic in the upper half-plane

$$D = \{z = x + iy : y > 0\}.$$

It turns out that the spaces $H^p(D)$ and $E^p(D)$, as defined in Chapter 10, do not coincide in this case, and in fact $E^p(D)$ is properly contained in $H^p(D)$. It is natural also to consider the space $\mathfrak{H}^p$ $(0 < p < \infty)$ of functions f analytic in D, such that $|f(x + iy)|^p$ is integrable for each $y > 0$ and

$$\mathfrak{M}_p(y, f) = \left\{ \int_{-\infty}^{\infty} |f(x + iy)|^p \, dx \right\}^{1/p}$$

is bounded, $0 < y < \infty$. $\mathfrak{H}^\infty$ will denote the space of bounded analytic functions in D. Eventually it will turn out that $\mathfrak{H}^p = E^p(D)$, but the general theory of Chapter 10 is not entirely applicable because the boundary of D is not rectifiable.

It is possible to develop the theory of $\mathfrak{H}^p$ by mapping the half-plane onto the unit disk, but this approach runs into difficulties because the lines $y = y_0$ are mapped onto circles tangent to the unit circle. Mainly to deal with this problem, or rather to avoid it, we begin with some lemmas on subharmonic functions in D. We then discuss boundary behavior, factorization, and integral representations of $\mathfrak{H}^p$ functions, basing most of the proofs on known properties of H^p functions in the disk. The chapter concludes with the Paley–Wiener theorem, in which a Fourier transform plays the role of the Taylor coefficients.

11.1. SUBHARMONIC FUNCTIONS

LEMMA 1. If $g(z) \geq 0$ is subharmonic in the upper half-plane D and

$$\int_{-\infty}^{\infty} g(x + iy) \, dx \leq M, \qquad y > 0,$$

then

$$g(z) \leq 4M/3\pi y, \qquad z = x + iy.$$

PROOF. Fix $z_0 = x_0 + iy_0$ $(y_0 > 0)$, and map D onto the unit disk by

$$w = \frac{z - z_0}{z - \bar{z}_0}.$$

Then

$$G(w) = g\left(\frac{z_0 - \bar{z}_0 w}{1 - w}\right)$$

is subharmonic in $|w| < 1$, so by the mean value theorem,

$$G(0) \leq \frac{1}{\pi\rho^2} \iint\limits_{|w|<\rho} G(w) \, du \, dv, \qquad \rho < 1,$$

where $w = u + iv$. Letting $\rho \to 1$, we therefore have

$$g(z_0) = G(0) \leq \frac{1}{\pi} \iint\limits_{|w|<1} G(w) \, du \, dv$$

$$= \frac{4y_0^2}{\pi} \iint\limits_{y>0} \frac{g(z)}{[(x - x_0)^2 + (y + y_0)^2]^2} \, dx \, dy$$

$$\leq \frac{4y_0^2}{\pi} \int_0^{\infty} \int_{-\infty}^{\infty} (y + y_0)^{-4} g(x + iy) \, dx \, dy \leq \frac{4M}{3\pi y_0}.$$

LEMMA 2. If a subharmonic function $g(z)$ satisfies the hypotheses of Lemma 1, then it has a harmonic majorant in D.

PROOF. Map D onto $|w| < 1$ by

$$w = \psi(z) = \frac{z - i}{z + i}; \qquad z = \varphi(w) = \frac{i(1 + w)}{1 - w}. \tag{1}$$

The line $y = b$ then corresponds to the circle C_b with center $b(1 + b)^{-1}$ and radius $R = (1 + b)^{-1}$. By Lemma 1, $G(w) = g(\varphi(w))$ is bounded inside C_b, so it has a least harmonic majorant $U_b(w)$ there. If $a < b$, it is clear that $U_a(w) \geq U_b(w)$ for each w inside C_b. Thus by Harnack's principle (see Ahlfors [2], p. 236) and a diagonalization argument,

$$\lim_{a \to 0} U_a(w) = U(w), \qquad |w| < 1;$$

and $U(w)$ is a harmonic majorant of $G(w)$ unless $U(w) \equiv \infty$. Hence $u(z) = U(\psi(z))$ is a harmonic majorant of $g(z)$ if $U(w) \not\equiv \infty$.

To show that $U(w) \not\equiv \infty$, let Γ_b be a circle concentric with C_b and having radius $\rho < R$. Let $V_b(w)$ be the Poisson integral of G over Γ_b. Then, in particular,

$$V_b\left(\frac{b}{1 + b}\right) = \frac{1}{2\pi\rho} \int_{\Gamma_b} G(w)|dw|. \tag{2}$$

But as $\rho \to R$, $V_b(w) \to U_b(w)$ inside C_b; so it follows from (2) and the bounded convergence theorem that

$$U_b\left(\frac{b}{1 + b}\right) = \frac{1}{2\pi R} \int_{C_b} G(w)|dw|$$

$$\leq \frac{2(1 + b)}{\pi} \int_{C_b} \frac{G(w)}{|1 - w|^2} |dw|$$

$$\leq \frac{1 + b}{\pi} \int_{-\infty}^{\infty} g(x + ib)\, dx \leq \frac{(1 + b)M}{\pi}.$$

As $b \to 0$, this shows $U(0) \leq M/\pi$. Thus $U(w) \not\equiv \infty$, and the proof is complete.

11.2. BOUNDARY BEHAVIOR

The following theorem is an immediate consequence of Lemma 2.

THEOREM 11.1. If $0 < p < \infty$ and $f \in \mathfrak{H}^p$, then $f \in H^p(D)$.

COROLLARY. If $f \in \mathfrak{H}^p$, then the boundary function

$$f(x) = \lim_{y \to 0} f(x + iy)$$

exists almost everywhere, $f \in L^p$, and

$$\int_{-\infty}^{\infty} \frac{\log|f(x)|}{1 + x^2} \, dx > -\infty. \tag{3}$$

PROOF OF COROLLARY. As in the proof of Theorem 10.3, the existence (more generally) of a nontangential limit $f(x) = \lim_{z \to x} f(z)$ follows from the fact that $F(w) = f(\varphi(w))$ is in H^p, where φ is the mapping (1). Fatou's lemma shows $|f(x)|^p$ is integrable over $(-\infty, \infty)$. Finally, (3) follows from the fact (Theorem 2.2) that

$$\int_{|w|=1} \log|F(w)| \, |dw| > -\infty.$$

It is also true that $f(x + iy)$ tends to $f(x)$ in the L^p mean. Before showing this, it is convenient to prove a Poisson integral representation for $\mathfrak{H}^1$ functions and a factorization theorem analogous to that of F. Riesz (Theorem 2.5).

THEOREM 11.2. If $f \in \mathfrak{H}^p$, $1 \le p \le \infty$, then

$$f(z) = \frac{1}{\pi} \int_{-\infty}^{\infty} \frac{y}{(x-t)^2 + y^2} f(t) \, dt, \qquad z = x + iy. \tag{4}$$

Conversely, if $h \in L^p$ $(1 \le p \le \infty)$ and

$$f(z) = \frac{1}{\pi} \int_{-\infty}^{\infty} \frac{y}{(x-t)^2 + y^2} h(t) \, dt$$

is analytic in D, then $f \in \mathfrak{H}^p$ and its boundary function $f(x) = h(x)$ a.e.

PROOF. Since $F(w) = f(\varphi(w))$ is in H^p $(1 \le p \le \infty)$, it has a Poisson representation (Theorem 3.1)

$$F(w) = \frac{1}{2\pi} \int_0^{2\pi} \mathrm{Re}\left\{\frac{e^{i\theta} + w}{e^{i\theta} - w}\right\} F(e^{i\theta}) \, d\theta.$$

This gives (4) after a change of variable and a straightforward calculation. The converse is obtained from Jensen's inequality.

COROLLARY. If $f \in \mathfrak{H}^p$, $1 \le p < \infty$, then

$$\lim_{y \to 0} \int_{-\infty}^{\infty} |f(x + iy)|^p \, dx = \int_{-\infty}^{\infty} |f(x)|^p \, dx. \tag{5}$$

PROOF. Applying Jensen's inequality to (4), we have

$$\int_{-\infty}^{\infty} |f(x + iy)|^p \, dx \le \int_{-\infty}^{\infty} |f(x)|^p \, dx, \qquad y > 0. \tag{6}$$

This together with Fatou's lemma gives the result.

THEOREM 11.3. If $f \in \mathfrak{H}^p$ $(0 < p \le \infty)$ and $f(z) \not\equiv 0$, then $f(z) = b(z)g(z)$, where g is a nonvanishing $\mathfrak{H}^p$ function with $|g(x)| = |f(x)|$ a.e., and

$$b(z) = \left(\frac{z - i}{z + i}\right)^m \prod_n \frac{|z_n^2 + 1|}{z_n^2 + 1} \cdot \frac{z - z_n}{z - \bar{z}_n} \tag{7}$$

is a Blaschke product for the upper half-plane. Here m is a nonnegative integer and z_n are the zeros $(z_n \ne i)$ of f in D, finite or infinite in number. Furthermore,

$$\sum_n \frac{y_n}{1 + |z_n|^2} < \infty, \qquad z_n = x_n + iy_n. \tag{8}$$

PROOF. According to Theorem 2.5,

$$f(\varphi(w)) = B(w)G(w),$$

where B is a Blaschke product in the disk and $G \in H^p$ has no zeros. If we define $b(z) = B(\psi(z))$, the expression (7) follows from the corresponding formula for B; and (8) is equivalent to $\sum (1 - |\psi(z_n)|) < \infty$. It remains to show that the nonvanishing function $g(z) = G(\psi(z))$ is in $\mathfrak{H}^p$. But by Theorem 11.2, or rather by its proof, $[g(z)]^p$ is the Poisson integral of its boundary function; hence $g^p \in \mathfrak{H}^1$.

COROLLARY 1. If $f \in \mathfrak{H}^p$, $0 < p < \infty$, then (5) holds.

PROOF. Since $g^p \in \mathfrak{H}^1$,

$$\int_{-\infty}^{\infty} |f(x + iy)|^p \, dx \le \int_{-\infty}^{\infty} |g(x + iy)|^p \, dx \to \int_{-\infty}^{\infty} |g(x)|^p \, dx = \int_{-\infty}^{\infty} |f(x)|^p \, dx.$$

COROLLARY 2. If $f \in \mathfrak{H}^p$, $0 < p < \infty$, then $f(z) \to 0$ as $z \to \infty$ within each half-plane $y \ge \delta > 0$.

PROOF. Since $|f(z)|^p \le |g(z)|^p$, it is enough to prove this for $\mathfrak{H}^1$ functions. But each $f \in \mathfrak{H}^1$ has a Poisson representation of the form (4). Given $\varepsilon > 0$, choose T large enough so that

$$\int_{-\infty}^{-T} |f(t)| \, dt + \int_{T}^{\infty} |f(t)| \, dt < \varepsilon.$$

Then (4) gives

$$|f(z)| \leq \frac{1}{\pi} \int_{-T}^{T} \frac{y}{(x-t)^2 + y^2} |f(t)| \, dt + \frac{\varepsilon}{\pi \delta}$$

$$= O\left(\frac{1}{|z|}\right) + \frac{\varepsilon}{\pi \delta} \qquad (|z| \to \infty).$$

It is now a short step to the theorem on mean convergence.

THEOREM 11.4. If $f \in \mathfrak{H}^p$ $(0 < p < \infty)$, then

$$\lim_{y \to 0} \int_{-\infty}^{\infty} |f(x + iy) - f(x)|^p \, dx = 0.$$

PROOF. Apply Corollary 1 above and Lemma 1 in Section 2.3.

A further application of the factorization theorem shows that $\mathfrak{M}_p(y, f)$ is a nonincreasing function of y if $f \in \mathfrak{H}^p$. This is expressed by the following theorem.

THEOREM 11.5. If $f \in \mathfrak{H}^p$ $(0 < p < \infty)$ and $0 < y_1 < y_2$, then

$$\mathfrak{M}_p(y_1, f) \geq \mathfrak{M}_p(y_2, f).$$

PROOF. Let $f_1(z) = f(z + iy_1)$. Then $f_1 \in \mathfrak{H}^p$, so it has the factorization $f_1 = b_1 g_1$ as in Theorem 11.3. But since $g_1{}^p \in \mathfrak{H}^1$, an application of (6) gives

$$\mathfrak{M}_p(y_2 - y_1, f_1) \leq \mathfrak{M}_p(y_2 - y_1, g_1) \leq \mathfrak{M}_p(0, g_1) = \mathfrak{M}_p(0, f_1),$$

which proves the theorem.

The hypothesis $f \in \mathfrak{H}^p$ is essential, as the example

$$f(z) = e^{-iz}(i + z)^{-2/p}$$

shows. Here

$$\mathfrak{M}_p{}^p(y, f) = \pi e^y(y + 1)^{-1},$$

which increases to infinity with y.

11.3. CANONICAL FACTORIZATION

The factorization $f = bg$ given in Theorem 11.3 can be refined as it was in the case of the disk (Theorem 2.8) to produce a canonical factorization for $\mathfrak{H}^p$ functions. The space $H^p(D)$ will be considered first.

THEOREM 11.6. Each function $f \in H^p(D)$, $0 < p < \infty$, has a unique factorization of the form

$$f(z) = e^{i\alpha z} b(z) s(z) G(z), \tag{9}$$

where $\alpha \geq 0$, $b(z)$ is a Blaschke product of the form (7),

$$s(z) = \exp\left\{i \int_{-\infty}^{\infty} \frac{1 + tz}{t - z} \, dv(t)\right\} \tag{10}$$

for some nondecreasing function $v(t)$ of bounded variation over $(-\infty, \infty)$ with $v'(t) = 0$ a.e., and

$$G(z) = e^{i\gamma} \exp\left\{\frac{1}{\pi i} \int_{-\infty}^{\infty} \frac{(1 + tz) \log \omega(t)}{(t - z)(1 + t^2)} \, dt\right\} \tag{11}$$

for some real number γ and some measurable function $\omega(t) \geq 0$ with

$$\int_{-\infty}^{\infty} \frac{\log \omega(t)}{1 + t^2} > -\infty \quad \text{and} \quad \int_{-\infty}^{\infty} \frac{[\omega(t)]^p}{1 + t^2} \, dt < \infty. \tag{12}$$

Conversely, if f has the form (9), where $\alpha \geq 0$, $b(z)$ is an arbitrary Blaschke product, and the functions $v(t)$ and $\omega(t)$ have the properties indicated, then $f \in H^p(D)$.

PROOF. If $f \in H^p(D)$, then by Theorem 2.8,

$$F(w) = f(\varphi(w)) = B(w)S(w)\Phi(w),$$

where $B(w)$ is a Blaschke product for the disk,

$$S(w) = \exp\left\{-\int_0^{2\pi} \frac{e^{i\theta} + w}{e^{i\theta} - w} \, d\mu(\theta)\right\},$$

$\mu(\theta)$ being a bounded nondecreasing singular function, and

$$\Phi(w) = e^{i\gamma} \exp\left\{\frac{1}{2\pi} \int_0^{2\pi} \frac{e^{i\theta} + w}{e^{i\theta} - w} \log|F(e^{i\theta})| \, d\theta\right\}.$$

As in the proof of Theorem 11.3, $b(z) = B(\psi(z))$ is a Blaschke product of the form (7). With $w = \psi(z)$ and $e^{i\theta} = \psi(t)$, a calculation gives

$$\frac{e^{i\theta} + w}{e^{i\theta} - w} = \frac{1 + tz}{i(t - z)}$$

and

$$d\theta = \frac{2dt}{1 + t^2}.$$

Thus $G(z) = \Phi(\psi(z))$ has the form (11), with $\omega(t) = |f(t)|$. The properties (12) of ω follow from the properties $\log|F(e^{i\theta})| \in L^1$ and $F(e^{i\theta}) \in L^p$. Finally, taking into account the possible jumps of μ at 0 and at 2π, we have

$$\log S(\psi(z)) = i \int_{-\infty}^{\infty} \frac{1 + tz}{t - z} \, dv(t) + i\alpha z,$$

where $v(t) = \mu(\arg\{\psi(t)\})$ and

$$\alpha = \mu(0+) - \mu(0) + \mu(2\pi) - \mu(2\pi-) \geq 0.$$

Conversely, if f is an arbitrary function of the form (9), then

$$|f(z)|^p \leq |G(z)|^p = \exp\left\{ \frac{1}{\pi} \int_{-\infty}^{\infty} \frac{y \log[\omega(t)]^p}{(x - t)^2 + y^2} \, dt \right\}$$

$$\leq \frac{1}{\pi} \int_{-\infty}^{\infty} \frac{y[\omega(t)]^p}{(x - t)^2 + y^2} \, dt, \qquad z = x + iy, \tag{13}$$

by the geometric–arithmetic mean inequality. Thus $|f(z)|^p$ has a harmonic majorant in D.

THEOREM 11.7. Each function $f \in \mathfrak{H}^p$, $0 < p < \infty$, has a unique factorization of the form (9), with the factors defined as in Theorem 11.6 except that the second condition in (12) is replaced by $\omega \in L^p$. Conversely, each product of such factors belongs to $\mathfrak{H}^p$.

PROOF. Since $f \in \mathfrak{H}^p$ implies $f \in H^p(D)$ and $f(t) \in L^p$, the first statement follows from Theorem 11.6. The converse is proved by integrating (13) with respect to x.

11.4. CAUCHY INTEGRALS

We now wish to show that every function in $\mathfrak{H}^p$ ($1 \leq p < \infty$) can be recovered from its boundary function by a Cauchy integral. It would be possible to prove this by mapping the half-plane onto the disk, as we did in Section 10.2 for the case of a domain bounded by a rectifiable Jordan curve. However, it is much easier to base the discussion on the identity

$$\frac{1}{t - z} - \frac{1}{t - \bar{z}} = \frac{2iy}{(x - t)^2 + y^2}, \qquad z = x + iy, \tag{14}$$

and essentially to follow the argument used for the disk in Section 3.3.

THEOREM 11.8. If $f \in \mathfrak{H}^p$ $(1 \leq p < \infty)$, then

$$f(z) = \frac{1}{2\pi i} \int_{-\infty}^{\infty} \frac{f(t)}{t - z} \, dt, \qquad \operatorname{Im}\{z\} = y > 0;$$

and the integral vanishes for all $y < 0$. Conversely, if $h \in L^p$ $(1 \leq p < \infty)$ and

$$\frac{1}{2\pi i} \int_{-\infty}^{\infty} \frac{h(t)}{t - z} \, dt \equiv 0, \qquad y < 0,$$

then for $y > 0$ this integral represents a function $f \in \mathfrak{H}^p$ whose boundary function $f(x) = h(x)$ a.e.

PROOF. If $f \in \mathfrak{H}^p$, the Cauchy integral

$$F(z) = \frac{1}{2\pi i} \int_{-\infty}^{\infty} \frac{f(t)}{t - z} \, dt$$

is analytic in both of the half-planes $y > 0$ and $y < 0$. According to (14), it is related to the Poisson integral by the identity

$$F(z) - F(\bar{z}) = \frac{1}{\pi} \int_{-\infty}^{\infty} \frac{y}{(x - t)^2 + y^2} f(t) \, dt.$$

Therefore, in view of Theorem 11.2,

$$F(\bar{z}) = F(z) - f(z), \qquad y > 0.$$

In particular, $F(\bar{z})$ is analytic for $y > 0$, so $F(z)$ must be identically constant in the lower half-plane. But since $F(z) \to 0$ as $z \to \infty$, the constant is zero. Thus $F(z) = f(z)$ in $y > 0$ and $F(z) = 0$ in $y < 0$, which was to be shown. The converse follows immediately from Theorem 11.2.

11.5. FOURIER TRANSFORMS

We come now to the Paley–Wiener theorem, a half-plane analogue of the fact that H^2 is the class of power series $\sum a_n z^n$ with $\sum |a_n|^2 < \infty$. For functions analytic in the upper half-plane, the Fourier integral

$$f(z) = \int_0^{\infty} e^{izt} F(t) \, dt \tag{15}$$

plays the role of a power series. Before stating the Paley–Wiener theorem, we recall a few facts about Fourier transforms of L^2 functions.

If $f \in L^2$, its Fourier transform is defined as

$$\hat{f}(x) = \underset{R \to \infty}{\text{l.i.m.}} \frac{1}{2\pi} \int_{-R}^{R} e^{-ixt} f(t) \, dt,$$

where "l.i.m." stands for "limit in mean" in the L^2 sense. It is a theorem of Plancherel that $\hat{f}$ exists, $\|f\|_2^2 = 2\pi\|\hat{f}\|_2^2$, and

$$f(t) = \underset{R \to \infty}{\text{l.i.m.}} \int_{-R}^{R} e^{ixt}\hat{f}(x)\, dx.$$

If g is another L^2 function with Fourier transform $\hat{g}$, the *Plancherel formula* is

$$\int_{-\infty}^{\infty} f(t)\hat{g}(t)\, dt = \int_{-\infty}^{\infty} \hat{f}(t)g(t)\, dt. \tag{16}$$

THEOREM 11.9 (Paley–Wiener). A function $f(z)$ belongs to $\mathfrak{H}^2$ if and only if it has the form (15) for some $F \in L^2$.

PROOF. If f has the form (15) with $F \in L^2$, it is analytic in the upper half-plane, as an application of Morera's theorem shows. For fixed $y > 0$, the function $f_y(x) = f(x + iy)$ is the inverse Fourier transform of

$$\hat{f}_y(t) = \begin{cases} e^{-yt}F(t), & t \geq 0 \\ 0, & t < 0. \end{cases}$$

Hence

$$\int_{-\infty}^{\infty} |f(x + iy)|^2\, dx = 2\pi \int_0^{\infty} e^{-2yt}|F(t)|^2\, dt$$

$$\leq 2\pi \int_0^{\infty} |F(t)|^2\, dt < \infty,$$

showing that $f \in \mathfrak{H}^2$.

Conversely, each f in $\mathfrak{H}^2$ is the Cauchy integral of its boundary function, by Theorem 11.8:

$$f(z) = \frac{1}{2\pi i} \int_{-\infty}^{\infty} \frac{f(t)}{t - z}\, dt, \qquad y > 0. \tag{17}$$

But

$$\frac{1}{2\pi i(t - z)} = \frac{1}{2\pi} \int_0^{\infty} e^{-it\xi}e^{iz\xi}\, d\xi = \hat{\sigma},$$

where $\sigma(\xi) = e^{iz\xi}$ for $\xi \geq 0$ and $\sigma(\xi) = 0$ for $\xi < 0$. Thus the Plancherel formula combined with (17) gives

$$f(z) = \int_{-\infty}^{\infty} \hat{f}(\xi)\sigma(\xi)\, d\xi = \int_0^{\infty} e^{iz\xi}\hat{f}(\xi)\, d\xi,$$

which proves the Paley–Wiener theorem.

COROLLARY. If $f \in \mathfrak{H}^2$ and $\hat{f}$ is the Fourier transform of its boundary function, then $\hat{f}(\xi) = 0$ for almost all $\xi < 0$.

PROOF. By Theorem 11.8, the Cauchy integral (17) vanishes for all $y < 0$. But if $y < 0$,

$$\frac{1}{2\pi i(z - t)} = \frac{1}{2\pi} \int_{-\infty}^{0} e^{-it\xi}e^{iz\xi} \, d\xi,$$

and we find as before that

$$\int_{-\infty}^{0} e^{iz\xi}\hat{f}(\xi) \, d\xi \equiv 0, \qquad y < 0.$$

In particular,

$$\int_{-\infty}^{0} e^{2\xi}|\hat{f}(\xi)|^2 \, d\xi = 0,$$

which proves $\hat{f}(\xi) = 0$ for almost all $\xi < 0$.

The argument can be generalized to give a similar representation for $\mathfrak{H}^p$ functions, $1 \leq p < 2$. We shall content ourselves with a discussion of $\mathfrak{H}^1$. If $f \in L^1$, its Fourier transform

$$\hat{f}(x) = \frac{1}{2\pi} \int_{-\infty}^{\infty} e^{-ixt}f(t) \, dt$$

is continuous on $-\infty < x < \infty$, and $\hat{f}(x) \to 0$ as $x \to \pm\infty$. If also $g \in L^1$, the formula (16) is a simple consequence of Fubini's theorem. Thus the proof of the Paley–Wiener theorem can be adapted to obtain the following result.

THEOREM 11.10. If $f \in \mathfrak{H}^1$ and $\hat{f}$ is the Fourier transform of the boundary function, then $\hat{f}(\xi) = 0$ for all $\xi \leq 0$ and

$$f(z) = \int_{0}^{\infty} e^{iz\xi}\hat{f}(\xi) \, d\xi, \qquad y > 0.$$

EXERCISES

1. Show that $f \in \mathfrak{H}^p$ if and only if $f(\varphi(w))[\varphi'(w)]^{1/p} \in H^p$, $0 < p < \infty$. Hence show that $\mathfrak{H}^p = E^p(D)$, where D is the upper half-plane. [*Suggestion*: Use the canonical factorization theorems (Theorems 11.6 and 11.7).]

2. Show that $E^p(D)$ is properly contained in $H^p(D)$ if D is the upper half-plane.

3. Give an example of a function $f(z)$ which is analytic in a half-plane $y > -\delta$ $(\delta > 0)$, with $f(x) \in L^1$, but which is not the Cauchy integral of $f(x)$.

4. Let $\mu(t)$ be a complex-valued function of bounded variation over $(-\infty, \infty)$, such that

$$\int_{-\infty}^{\infty} e^{ixt} \, d\mu(t) = 0 \qquad \text{for all} \quad x > 0.$$

Show that $d\mu$ is absolutely continuous with respect to Lebesgue measure.

5. Prove the half-plane analogue of Hardy's inequality (Section 3.6): If $f \in \mathfrak{H}^1$ and $\hat{f}$ is the Fourier transform of its boundary function, then

$$\int_0^{\infty} \frac{|\hat{f}(t)|}{t} \, dt \le \tfrac{1}{2} \int_{-\infty}^{\infty} |f(x)| \, dx.$$

(Hille and Tamarkin [3]. See Exercise 4 of Chapter 3.)

6. For $f \in \mathfrak{H}^p$, $0 < p < \infty$, prove

$$\int_0^{\infty} |f(x + iy)|^p \, dy \le \tfrac{1}{2} \int_{-\infty}^{\infty} |f(x)|^p \, dx.$$

(This analogue of the Fejér–Riesz theorem is due to M. Riesz [1].)

7. Show that if $b(z)$ is a Blaschke product in the upper half-plane, then

$$\lim_{y \to 0} \int_{-\infty}^{\infty} \frac{\log|b(x + iy)|}{1 + x^2} \, dx = 0.$$

Conversely, show that if $f(z)$ is analytic in $y > 0$, $|f(z)| < 1$, and

$$\lim_{y \to 0} \int_{-\infty}^{\infty} \frac{\log|f(x + iy)|}{1 + x^2} \, dx = 0,$$

then $f(z) = e^{i(\gamma + \alpha z)} b(z)$, where γ is a real number, $\alpha \ge 0$, and $b(z)$ is a Blaschke product (Akutowicz [1]).

NOTES

Most of the results in Section 11.2 are due to Hille and Tamarkin [3], who considered only $1 \le p < \infty$. They proved the key result $\mathfrak{H}^p \subset H^p(D)$ using a lemma of Gabriel [2] on subharmonic functions. But in order to apply Gabriel's lemma, it must first be shown that a function $f \in \mathfrak{H}^p$ tends to a limit as $z \to \infty$ within each half-plane $y \ge \delta > 0$. Hille and Tamarkin were able to show this only after a difficult argument proving the Poisson representation

(Theorem 11.2) from first principles. Kawata [1] extended the Hille–Tamarkin results to $0 < p < 1$. The relatively simple approach via harmonic majorants, as presented in the text, is due to Krylov [1]. Krylov also obtained the canonical factorization theorems of Section 11.3. Theorem 11.9 is in the book of Paley and Wiener [1]. For proofs of the Plancherel theorems and other information about Fourier transforms, see Goldberg [1]. Theorem 11.10 is due to Hille and Tamarkin [1]; see their papers [2, 3] for further results. Kawata [1] proved theorems on the growth of $\mathfrak{M}_p(y, f)$ analogous to those of Hardy and Littlewood for the disk.

The purpose of this final chapter is to give a self-contained proof of the "corona theorem," which concerns the maximal ideal space of the Banach algebra H^∞. After describing the result in its abstract form, we show how it reduces to a certain "concrete" theorem. Here the discussion must presuppose an elementary acquaintance with the theory of Banach algebras. However, the proof of the reduced theorem (which occupies most of the chapter) uses purely classical methods, and makes no further reference to Banach algebras.

12.1. MAXIMAL IDEALS

Let A be a commutative Banach algebra with unit, and let $\mathcal{M}$ be its maximal ideal space, endowed with the Gelfand topology. In other words, the basic neighborhoods of a point $M^* \in \mathcal{M}$ have the form

$$\mathcal{U} = \{M \in \mathcal{M} : |\hat{x}_k(M) - \hat{x}_k(M^*)| < \varepsilon, \quad k = 1, \ldots, n\},$$

where $\varepsilon > 0$, the x_k are arbitrary elements of A, and $\hat{x}_k$ is the Gelfand transform of x_k. That is, $\hat{x}(M) = \phi_M(x)$, where ϕ_M is the multiplicative linear functional with kernel M. Now let

$$y_k = x_k - \hat{x}_k(M^*)e,$$

where e is the unit element of A. Then $y_k \in M^*$ (since $\hat{y}_k(M^*) = 0$), and $\mathscr{U}$ takes the equivalent form

$$\mathscr{U} = \{M \in \mathscr{M} : |\hat{y}_k(M)| < \varepsilon, \quad k = 1, \ldots, n\}.$$

It is well known that $\mathscr{M}$ is a compact Hausdorff space under the Gelfand topology.

Associated with each fixed point ζ, $|\zeta| < 1$, the Banach algebra H^∞ has the maximal ideal

$$M_\zeta = \{f \in H^\infty : f(\zeta) = 0\}.$$

The problem arises to describe the closure of these ideals M_ζ in the maximal ideal space $\mathscr{M}$ of H^∞, under the Gelfand topology. Are there points in $\mathscr{M}$ which are outside this closure? To put the question in more picturesque language, does the unit disk have a "corona"? As it turns out, the answer is negative.

CORONA THEOREM. The maximal ideals M_ζ, $|\zeta| < 1$, are dense in the maximal ideal space of H^∞.

The corona theorem is a direct consequence of the following purely function-theoretic result.

THEOREM 12.1 (Reduced corona theorem). Let $f_1, f_2, \ldots, f_n$ be functions in H^∞ such that

$$|f_1(z)| + |f_2(z)| + \cdots + |f_n(z)| \geq \delta, \qquad |z| < 1,$$

for some $\delta > 0$. Then there are functions $g_1, g_2, \ldots, g_n$ in H^∞ such that

$$f_1(z)g_1(z) + f_2(z)g_2(z) + \cdots + f_n(z)g_n(z) \equiv 1.$$

To derive the corona theorem from Theorem 12.1, suppose the maximal ideals M_ζ are *not* dense in $\mathscr{M}$. Then some $M^* \in \mathscr{M}$ has a neighborhood of the form

$$\mathscr{U} = \{M \in \mathscr{M} : |\hat{f}_k(M)| < \varepsilon, \quad k = 1, \ldots, n\}, \qquad f_k \in M^*,$$

which contains no ideal M_ζ. In other words, to each point ζ ($|\zeta| < 1$) there corresponds an integer k ($k = 1, \ldots, n$) such that

$$|f_k(\zeta)| = |\hat{f}_k(M_\zeta)| \geq \varepsilon.$$

In particular, $|f_1(\zeta)| + \cdots + |f_n(\zeta)| \geq \varepsilon$ for all ζ, $|\zeta| < 1$. But by Theorem 12.1, this implies the existence of $g_1, \ldots, g_n$ in H^∞ such that $f_1 g_1 + \cdots + f_n g_n = e$.

Since $f_k \in M^*$ $(k = 1, \ldots, n)$, it then follows that $e \in M^*$, which is impossible. Thus the ideals M_ζ are dense in $\mathcal{M}$, as the corona theorem asserts.

The rest of the chapter will be devoted to the proof of Theorem 12.1.

12.2. INTERPOLATION AND THE CORONA THEOREM

The most difficult step in the proof of the corona theorem is to show that the zeros of an arbitrary finite Blaschke product can be surrounded by a contour which is not "too long," and on which the Blaschke product is neither "too large" nor "too small". That the contour is not too long will mean that the arclength measure it induces on the unit disk is a Carleson measure, in the sense of Section 9.5. The precise statement is as follows.

LEMMA 1 (Carleson's lemma). There exist absolute constants κ $(0 < \kappa < 1)$ and $C > 0$ for which the following is true. Corresponding to each ε $(0 < \varepsilon \leq \frac{1}{4})$ and to each finite Blaschke product

$$B(z) = \prod_{v=1}^{s} \frac{z - a_v}{1 - \overline{a}_v z}, \qquad |a_v| < 1, \tag{1}$$

there is a rectifiable contour Γ such that
   (i) Γ has winding number 1 about each point a_v;
   (ii) $\varepsilon \leq |B(z)| \leq \varepsilon^\kappa$ for all $z \in \Gamma$;
   (iii) the measure μ defined on $|z| < 1$ by letting $\mu(E)$ be the arclength of that part of Γ which lies in the set E, has the property

$$\mu(S) \leq C\varepsilon^{-2}h$$

for each set S of the form

$$S = \{z = re^{i\theta} : 1 - h \leq r < 1; \quad \theta_0 \leq \theta \leq \theta_0 + h\}.$$

Since the proof of this lemma is long and technical, we shall defer it to the end of the chapter (Sections 12.3–12.5). It seems advantageous first to motivate the lemma by showing how it leads to a proof of the (reduced) corona theorem, by way of the following result on interpolation.

LEMMA 2. Let $B(z)$ be a finite Blaschke product of the form (1) with distinct zeros a_v $(v = 1, \ldots, s)$. For $\delta < \frac{1}{2}$, let $F(z)$ be analytic in the (possibly disconnected) set $\{z : |B(z)| < \delta\}$ and satisfy $|F(z)| < 1$ there. Then there exists $f \in H^\infty$ with $f(a_v) = F(a_v)$, $v = 1, \ldots, s$; and $\|f\| \leq \delta^{-\alpha}$, where α is an absolute constant.

PROOF. Obviously, there are many functions $f \in H^\infty$ with $f(a_v) = F(a_v)$, $v = 1, \ldots, s$. Choose such an interpolating function f with minimal norm. Then by the duality relation [see Section 8.5, example (ii)],

$$\|f\|_\infty = \left| \sum_{v=1}^{s} \frac{g(a_v)F(a_v)}{B'(a_v)} \right|$$

for some $g \in H^1$ of norm $\|g\|_1 = 1$. With κ as in Carleson's lemma, choose ε such that $\varepsilon^\kappa = \delta/2$, and let Γ be the corresponding contour. Then we have

$$\|f\|_\infty = \left| \frac{1}{2\pi i} \int_\Gamma \frac{g(z)F(z)}{B(z)} \, dz \right|$$

$$\leq \frac{1}{2\pi\varepsilon} \int_\Gamma |g(z)| \, |dz|.$$

But the arclength measure μ induced by Γ is a Carleson measure [Lemma 1, (iii)], so by Carleson's theorem (Theorem 9.3 and the remark following the proof), there is an absolute constant c such that

$$\|f\|_\infty \leq c\varepsilon^{-5} \|g\|_1 = c\varepsilon^{-5} \leq \delta^{-\alpha},$$

where α is an absolute constant. This proves Lemma 2.

In using Lemma 2 to prove the corona theorem, one must approximate certain functions by finite Blaschke products. The following lemma is slightly stronger than what is actually needed.

LEMMA 3. Let $f(z)$ be analytic in the open unit disk D and continuous in $\bar{D}$. Suppose $0 < |f(z)| \leq 1$ on $|z| = 1$ and let E be the subset of $|z| = 1$ on which $|f(z)| < 1$. Suppose E is nonempty. Then there exists a sequence $\{B_n(z)\}$ of finite Blaschke products with simple zeros, such that $|B_n(z)| \to |f(z)|$ uniformly in each closed subset of $(\bar{D} - \bar{E})$, and $B_n(z) \to f(z)$ uniformly in each closed subset of D.

PROOF. Let S be an arbitrary closed subset of $(\bar{D} - \bar{E})$. For $0 < \rho < 1$, let $f_\rho(z) = f(\rho z)$, and let E_ρ be the subset of $|z| = 1$ for which $|f_\rho(z)| < 1$. Then $f_\rho(z) \to f(z)$ uniformly in $\bar{D}$ as $\rho \to 1$, and for ρ sufficiently near 1, $S \subset (\bar{D} - \bar{E}_\rho)$. Hence it will suffice to prove the lemma under the assumption that f is analytic in $\bar{D}$, and therefore has at most a finite number of zeros in D. Then, since it is clear that a finite Blaschke product can be approximated by one with simple zeros, uniformly in $\bar{D}$, it is enough to suppose that f does not vanish in $\bar{D}$. Assuming for convenience that $f(0) > 0$, we then have

$$f(z) = \exp\left\{ \frac{1}{2\pi} \int_0^{2\pi} \frac{e^{it} + z}{e^{it} - z} \log |f(e^{it})| \, dt \right\}.$$

Now let $\omega_k = e^{2\pi i k/n}$ be the nth roots of unity $(k = 1, \ldots, n)$, and let

$$f_n(z) = \exp\left\{\frac{1}{n} \sum_{k=1}^{n} \frac{\omega_k + z}{\omega_k - z} \log |f(\omega_k)|\right\}.$$

Elementary considerations show that $f_n(z) \to f(z)$ uniformly in S. Let

$$\varepsilon_k = -\frac{1}{n} \log |f(\omega_k)|,$$

so that

$$0 \le \varepsilon_k \le -\frac{1}{n} \log \mu = \delta_n,$$

say, where μ is the minimum of $|f(z)|$ on $|z| = 1$. Choosing n so large that $\delta_n < \frac{1}{2}$, let

$$1 - \rho_k^2 = 2\varepsilon_k \qquad \text{and} \qquad a_k = \rho_k \omega_k;$$

and define

$$B_n(z) = \prod_{k=1}^{n} \frac{\overline{a_k}}{|a_k|} \frac{a_k - z}{1 - \overline{a_k} z}.$$

Note that $|a_k| = 1$ if $\varepsilon_k = 0$, so that the corresponding factor in $B_n(z)$ is trivial. A calculation gives

$$2 \log |B_n(z)| = -2(1 - |z|^2) \sum_{k=1}^{n} \varepsilon_k |1 - \overline{a_k} z|^{-2} + O(\delta_n^2),$$

uniformly in S. From this it follows that

$$\log |B_n(z)| = \log |f_n(z)| + O(\delta_n),$$

uniformly in S. Hence $\log |B_n(z)| \to \log |f(z)|$, which implies $|B_n(z)| \to |f(z)|$, uniformly in S. Since $B_n(0) > 0$, it also follows (by analytic completion of the Poisson formula) that $B_n(z) \to f(z)$ uniformly in each disk $|z| \le r_0 < 1$.

Using Lemmas 2 and 3, we can now carry out the proof of the corona theorem. In fact, the argument will give Theorem 12.1 in the following sharper form.

THEOREM 12.2. Let $f_1, \ldots, f_n$ be H^∞ functions with $\|f_k\| \le 1$ $(k = 1, \ldots, n)$ and

$$|f_1(z)| + \cdots + |f_n(z)| \ge \delta, \qquad |z| < 1, \tag{2}$$

where $0 < \delta < \frac{1}{2}$. Then there exist functions $g_1, \ldots, g_n$ in H^∞ such that

$$f_1(z)g_1(z) + \cdots + f_n(z)g_n(z) \equiv 1, \qquad |z| < 1, \tag{3}$$

and $\|g_k\| \le \delta^{-\beta_n}$, where β_n is a constant depending only on n.

PROOF. The argument will proceed by induction on n. The theorem is trivial for $n = 1$. Suppose it has been proved for all collections of $(n-1)$ functions, and let $f_1, \ldots, f_n$ satisfy the given hypotheses. Suppose first that f_n is a finite Blaschke product:

$$f_n(z) = B(z) = \prod_{v=1}^{s} \frac{z - a_v}{1 - \bar{a}_v z},$$

with distinct a_v, $0 \le |a_v| < 1$. Then

$$S_\delta = \left\{ z : |B(z)| < \frac{\delta}{2} \right\} = \bigcup_{j=1}^{m} D_j,$$

where the D_j are the simply connected components of S_δ. Since

$$|f_1(z)| + \cdots + |f_{n-1}(z)| \ge \delta/2, \qquad z \in S_\delta,$$

and since the statement of the theorem is conformally invariant, it follows from the inductive hypothesis that in each domain D_j there exist $(n-1)$ bounded analytic functions G_{jk} with

$$|G_{jk}(z)| \le \left(\frac{2}{\delta} \right)^{\beta_{n-1}}, \qquad k = 1, \ldots, n-1,$$

and

$$\sum_{k=1}^{n-1} f_k(z) G_{jk}(z) = 1$$

for all z in D_j. By Lemma 2, there exist functions $g_1, \ldots, g_{n-1}$ in H^∞ such that

$$g_k(a_v) = G_{jk}(a_v), \qquad a_v \in D_j, \quad j = 1, \ldots, m,$$

and

$$\|g_k\| \le \left(\frac{2}{\delta} \right)^{\alpha + \beta_{n-1}}, \qquad k = 1, \ldots, n-1.$$

Now define

$$g_n(z) = [B(z)]^{-1} \left[1 - \sum_{k=1}^{n-1} f_k(z) g_k(z) \right].$$

This function is analytic in $|z| < 1$, since

$$\sum_{k=1}^{n-1} f_k(a_v) g_k(a_v) = \sum_{k=1}^{n-1} f_k(a_v) G_{jk}(a_v) = 1, \qquad a_v \in D_j.$$

The relation (3) is automatically satisfied, and for all $z \notin S_\delta$,

$$|g_n(z)| \le \left(\frac{2}{\delta} \right) \left[1 + (n-1) \left(\frac{2}{\delta} \right)^{\alpha + \beta_{n-1}} \right] \le \delta^{-\gamma_n},$$

where γ_n depends only on n. By the maximum modulus theorem, the same estimate holds throughout $|z| < 1$. This concludes the proof for the case in which f_n is a finite Blaschke product with simple zeros.

Continuing the inductive argument, suppose next that, more generally, $f_n(z)$ is analytic in $|z| \le 1$ and $f_n(z) \ne 0$ on $|z| = 1$. Let $F_n(z)$ be analytic and nonvanishing in $|z| < 1$, continuous in $|z| \le 1$, and satisfy

$$|F_n(z)| = \min\{|f_n(z)|^{-1}, 3/\delta\} \qquad \text{on} \quad |z| = 1.$$

Then $\Phi_n = F_n f_n$ has the property $|\Phi_n(z)| \le 1$ in $|z| \le 1$, and $|\Phi_n(z)| = 1$ on $|z| = 1$ except on the set where $|f_n(z)| < \delta/3$. Thus by Lemma 3, there is a sequence $\{B_m(z)\}$ of finite Blaschke products with simple zeros, such that $B_m(z) \to \Phi_n(z)$ in $|z| < 1$ as $m \to \infty$; and

$$|B_m(z)| \ge |\Phi_n(z)| - \delta/2, \qquad m \ge m_0, \tag{4}$$

uniformly on the set where $|f_n(z)| \ge \delta/2$. Since $|F_n(z)| \ge 1$ in $|z| \le 1$, it now follows from (2) and (4) that

$$|f_1(z)| + \cdots + |f_{n-1}(z)| + |B_m(z)| \ge \delta/2, \qquad |z| < 1, \quad m \ge m_0.$$

Thus by what has already been proved, there are H^∞ functions $g_{1m}, \dots,$ g_{nm} $(m \ge m_0)$ such that

$$f_1(z)g_{1m}(z) + \cdots + f_{n-1}(z)g_{n-1,m}(z) + B_m(z)g_{nm}(z) = 1$$

in $|z| < 1$ and

$$\|g_{km}\| \le \left(\frac{2}{\delta}\right)^{\gamma_n} \le \delta^{-\beta_n},$$

for a sufficiently large choice of β_n. Now choose a sequence $\{m_i\}$ such that $g_{km_i}(z) \to g_k(z)$ uniformly in each disk $|z| \le r_0 < 1$. Then $g_k \in H^\infty$, $\|g_k\| \le \delta^{-\beta_n}$, and (3) holds.

Finally, if f_n is an arbitrary H^∞ function, we may choose $\rho < 1$ such that $f_n(z) \ne 0$ on $|z| = \rho$. Then by what has just been proved, there are functions $g_k^{(\rho)} \in H^\infty$ with $\|g_k^{(\rho)}\| \le \delta^{-\beta_n}$ and

$$f_1(\rho z)g_1^{(\rho)}(z) + \cdots + f_n(\rho z)g_n^{(\rho)}(z) \equiv 1.$$

Now let $\rho \to 1$ through a suitable sequence, so that $g_k^{(\rho)}$ tends pointwise to $g_k \in H^\infty$, $k = 1, \dots, n$. This concludes the proof.

12.3. HARMONIC MEASURES

In order to complete the proof of the corona theorem, it now remains only to verify Carleson's lemma. This we shall do in several stages. In the present section, we digress to establish a general result ("Hall's lemma") on the harmonic measures of certain sets, and to apply it to obtain a special estimate which will enter into the argument at a later stage (Section 12.5). Hall's lemma also has more direct applications to function theory, but these lie outside the scope of the present discussion.

Let E be a closed subset of the right half-plane

$$H = \{z : \operatorname{Re}\{z\} > 0\}.$$

Suppose E does not divide the plane, and that the Dirichlet problem is solvable for $D = H - E$. Let ∂E denote the boundary of E, and let

$$E^* = \{i|z| : z \in E\}$$

be the circular projection of E onto the positive imaginary axis. Let $\omega(z)$ be the harmonic measure of E with respect to D. In other words, $\omega(z)$ is the bounded harmonic function in D for which $\omega(iy) = 0$ $(-\infty < y < \infty)$ and $\omega(z) = 1$ for $z \in \partial E$. Finally, let

$$\omega^*(z) = \frac{1}{\pi} \int_{E^*} \frac{x \, dt}{x^2 + (y - t)^2}, \qquad z = x + iy,$$

be the harmonic measure of E^* with respect to H.

LEMMA 4 (Hall's Lemma). For $(x + iy) \in D$,

$$\omega(x + iy) \geq \tfrac{2}{3}\omega^*(x - i|y|). \tag{5}$$

PROOF. Suppose first that E consists of a finite number of radial segments

$$\{re^{i\theta_k} : a_k \leq r \leq b_k\}, \qquad k = 1, 2, \ldots, n; \quad |\theta_k| < \pi/2,$$

with the intervals (a_k, b_k) disjoint. Let

$$G(z, \zeta) = \log \left| \frac{z + \bar{\zeta}}{z - \zeta} \right|$$

denote the Green's function of H, and consider the function

$$U(z) = \frac{1}{2\pi} \int_E \frac{1}{\zeta} G(z, \zeta) \, ds, \qquad \zeta = \xi + i\eta,$$

where ds is the element of arclength on E. We claim that

$$\omega^*(x) \leq U(x), \qquad x > 0; \tag{6}$$

and

$$U(z) < \tfrac{3}{2}, \qquad \operatorname{Re}\{z\} > 0. \tag{7}$$

To prove (6), observe that on the semicircle $|\zeta| = \rho$, $\text{Re}\{\zeta\} \geq 0$, the function $\xi^{-1}G(x, \zeta)$ attains its minimum for $\xi = 0$; hence

$$\frac{1}{\xi} G(x, \zeta) \geq \frac{2x}{x^2 + \rho^2}, \qquad |\zeta| = \rho.$$

But this gives

$$U(x) \geq \frac{1}{\pi} \int_{E*} \frac{x}{x^2 + \rho^2} \, d\rho = \omega^*(x).$$

The proof of (7) is somewhat more difficult. Fix $z = x + iy$, and let $M(\rho)$ be the maximum of $\xi^{-1}G(z, \zeta)$ over the part of the circle $|\zeta - z| = \rho$ where $\text{Re}\{\zeta\} \geq 0$. Since on this circle

$$\frac{1}{\xi} G(z, \zeta) = \frac{1}{2\xi} \log\left(1 + \frac{4x\xi}{\rho^2}\right)$$

is a decreasing function of ξ, the maximum occurs for the smallest possible value of ξ. Thus

$$M(\rho) = \begin{cases} \dfrac{1}{x - \rho} \log\left(\dfrac{2x}{\rho} - 1\right), & x > \rho \\[3mm] \dfrac{2x}{\rho^2}, & x \leq \rho. \end{cases}$$

Now let $\psi(\rho)$ denote the total length of the part of E which lies in the disk $|\zeta - z| < \rho$. Since $\psi(\rho) \leq 2\rho$ and $M(\rho)$ is a decreasing function, integration by parts gives

$$U(z) \leq \frac{1}{2\pi} \int_0^\infty M(\rho) \, d\psi(\rho) = -\frac{1}{2\pi} \int_0^\infty \psi(\rho) \, dM(\rho)$$

$$\leq -\frac{1}{\pi} \int_0^\infty \rho \, dM(\rho) = \frac{1}{\pi} \int_0^\infty M(\rho) \, d\rho = \frac{\pi}{4} + \frac{2}{\pi} < \frac{3}{2}.$$

From (6) and (7), it is easy to deduce (5). Indeed, it follows from (7) that

$$\tfrac{3}{2}\omega(z) - U(z) \geq 0$$

on the boundary of D, so by the maximum principle, the same is true in D. This and (6) show that the function

$$\varphi(z) = \tfrac{3}{2}\omega(z) - \omega^*(z)$$

is non-negative on the positive real axis; while $\varphi(iy) = 0$ for $y < 0$, and $\varphi(z) \geq \tfrac{1}{2}$ for $z \in E$. By the maximum principle, then,

$$\omega(x + iy) \geq \tfrac{2}{3}\omega^*(x + iy), \qquad x > 0, \quad y < 0.$$

Hence by symmetry,

$$\omega(x + iy) \geq \tfrac{2}{3}\omega^*(x - iy), \qquad x > 0, \quad y > 0.$$

This proves the lemma for the special case in which E is a union of radial segments with nonoverlapping projections.

For a general compact set E, choose $\varepsilon > 0$ and consider the set

$$S_\varepsilon = \{z : \omega(z) > 1 - \varepsilon\}.$$

Clearly, $\partial E \subset S_\varepsilon$. Choose a set $\tilde{E} \subset S_\varepsilon$ which consists of a finite number of radial segments with nonoverlapping projections, for which $\tilde{E}^* = E^*$. (To see that this is possible, cover E by open disks in S_ε and apply the Heine–Borel theorem.) Let $\tilde{\omega}(z)$ be the harmonic measure of $\tilde{E}$. By what has just been proved,

$$\tilde{\omega}(x + iy) \geq \tfrac{2}{3}\omega^*(x - i|y|),$$

since $\tilde{E}^* = E^*$. But the function $[\omega(z) - \tilde{\omega}(z)]$ vanishes on the imaginary axis, is ≥ 0 on ∂E, and is $\geq -\varepsilon$ wherever it is defined on $\tilde{E}$. Thus by the maximum principle,

$$\omega(x + iy) + \varepsilon \geq \tilde{\omega}(x + iy) \geq \tfrac{2}{3}\omega^*(x - |iy|)$$

for $(x + iy) \in D$. Now let $\varepsilon \to 0$, and the lemma is proved for compact sets E.

Finally, suppose E is closed but unbounded. Let E_r be the intersection of E with the disk $|z| \leq r$, let E_r^* be its circular projection, and let $\omega_r(z)$ and $\omega_r^*(z)$ denote the respective harmonic measures. Then

$$\omega(x + iy) \geq \omega_r(x + iy) \geq \tfrac{2}{3}\omega_r^*(x - i|y|)$$

for each point $(x + iy) \in D$. But it is clear from the integral representations that $\omega_r^*(z) \to \omega^*(z)$ pointwise as $r \to \infty$. This completes the proof.

Hall's lemma will now be applied to obtain a special result needed in the proof of Carleson's lemma. Let R be the annulus $\rho < |z| < 1$, and let E_1 be a closed subset of R which does not divide the plane. Let $\omega_1(z)$ be the harmonic measure of E_1 with respect to $(R - E_1)$, and let

$$E_1^* = \{e^{i\theta} : re^{i\theta} \in E_1\}$$

be the radial projection of E_1 onto the outer boundary of R. For fixed $\beta < \pi/|\log \rho|$, let F_1^* be the part of E_1^* such that $|\theta| \leq \beta|\log \rho|$. Then the total length $|F_1^*|$ of F_1^* can be estimated as follows.

LEMMA 5. If $\rho^{1/3} \notin E_1$,

$$|F_1^*| \leq |\log \rho|[\cosh(\pi\beta/2)]^2\omega_1(\rho^{1/3}).$$

PROOF. The multiple-valued function

$$\zeta = \xi + i\eta = iz^{-i\pi/\log \rho} = e^{\pi\theta/\log \rho}\exp\left\{i\pi\left(\frac{1}{2} - \frac{\log r}{\log \rho}\right)\right\}$$

maps R onto the right half-plane $\operatorname{Re}\{\zeta\} > 0$. Let $z = \varphi(\zeta)$ denote the (single-valued) inverse, and let

$$E = \{\zeta : \varphi(\zeta) \in E_1\}.$$

Then E is a closed (unbounded) subset of H, and Hall's lemma may be applied. The harmonic measure of E with respect to $D = H - E$ is $\omega(\zeta) = \omega_1(\varphi(\zeta))$. Since φ maps the circular projection E^* of E onto E_1^*, the harmonic measure of E^* with respect to H is $\omega^*(\zeta) = \omega_1^*(\varphi(\zeta))$, where $\omega_1^*(z)$ is the harmonic measure of E_1^* with respect to R. Thus Hall's lemma gives

$$\omega_1(\varphi(\xi + i\eta)) \geq \tfrac{2}{3}\omega_1^*(\varphi(\xi - i|\eta|)), \qquad \xi + i\eta \in D.$$

Choosing $\eta \geq 0$, we obtain in particular

$$\omega_1(z) \geq \tfrac{2}{3}\omega_1^*(\rho/\bar{z}), \qquad z \in (R - E_1); \quad |z| \geq \rho^{1/2}. \tag{8}$$

On the other hand, if F^* is the image of F_1^* under the restriction of φ^{-1} to $|\theta| < \pi$, we have

$$\omega_1^*(\rho^{2/3}) = \omega^*(e^{-i\pi/6}) \geq \frac{2}{\pi}\int_{F^*}\frac{d\eta}{3 + (1 + 2\eta)^2}$$

$$= 2|\log \rho|^{-1}\int_{F_1^*}[e^{\pi\theta/\log \rho} + 1 + e^{-\pi\theta/\log \rho}]^{-1}\,d\theta$$

$$\geq 2|\log \rho|^{-1}[\cosh(\pi\beta/2)]^{-2}|F_1^*|.$$

This together with (8) proves Lemma 5.

12.4. CONSTRUCTION OF THE CONTOUR Γ

We are now ready to prove Carleson's lemma (Lemma 1), the key to the proof of the corona theorem.

Let $B(z)$ be a finite Blaschke product, suppose $0 < \varepsilon < 1$, and consider the sets

$$\mathscr{A}(\varepsilon) = \{z : |B(z)| \leq \varepsilon\},$$
$$\mathscr{B}(\varepsilon) = \{z : |z| \leq 1, |B(z)| \geq \varepsilon\}.$$

For $n = 0, 1, \ldots$ and $k = 1, 2, \ldots, 2^{n+1}$, let

$$R_{nk} = \{re^{i\theta} : 1 - 2^{-n} \leq r \leq 1 - 2^{-n-1}, \quad (k-1)2^{-n}\pi \leq \theta \leq k2^{-n}\pi\}.$$

Let N be a positive integer (to be chosen later), and subdivide R_{nk} into 2^{2N} parts by means of the radial lines

$$\theta = (k - 1 + j2^{-N})2^{-n}\pi, \qquad j = 1, \ldots, 2^N - 1,$$

and the circular arcs

$$r = 1 - 2^{-n} + j2^{-N-n-1}, \qquad j = 1, \ldots, 2^N - 1.$$

Denote these parts (boundaries included) by $R_{nk}(i)$, $i = 1, \ldots, 2^{2N}$. The actual numbering scheme is not important. The sets $R_{nk}(i)$ will be called the *blocks* of the *rectangle* R_{nk}.

Now fix $\varepsilon \leq \frac{1}{4}$ and let $\mathscr{A}_0$ be the union of all the blocks $R_{nk}(i)$ which meet $\mathscr{A}(\varepsilon)$. Thus $\mathscr{A}(\varepsilon) \subset \mathscr{A}_0$.

LEMMA 6. If $2^{-N} \leq \varepsilon/8$, then $\mathscr{A}_0 \subset \mathscr{A}(2\varepsilon)$.

PROOF. If $z_0 \in \mathscr{A}_0$, then z_0 is in some block $R_{nk}(i)$ which contains a point $z_1 \in \mathscr{A}(\varepsilon)$. Then

$$|B(z_0)| \leq |B(z_0) - B(z_1)| + \varepsilon.$$

But it follows easily from the Cauchy formula that $|B'(z)| \leq (1 - |z|^2)^{-1}$, so

$$|B(z_0) - B(z_1)| \leq \int_{z_0}^{z_1} |B'(z)| \, |dz| \leq |z_0 - z_1| \max_{z \in R_{nk}(i)} (1 - |z|^2)^{-1}$$

$$\leq (1 + 2\pi)2^{-N-n}2^n < 8 \cdot 2^{-N} \leq \varepsilon.$$

Thus $|B(z_0)| < 2\varepsilon$, which shows that $z_0 \in \mathscr{A}(2\varepsilon)$. This proves Lemma 6.

From now on, let N be the smallest integer such that $2^{-N} \leq \varepsilon/8$. (In particular, $N \geq 5$.) For fixed κ ($0 < \kappa < \frac{1}{2}$), let $\mathscr{B}_0 = \mathscr{B}(\varepsilon^\kappa)$. Since $\varepsilon \leq \frac{1}{4}$ and $\kappa < \frac{1}{2}$, it is easily seen that $\varepsilon^\kappa > 2\varepsilon$. Hence by Lemma 6, $\mathscr{A}_0 \cap \mathscr{B}_0 = \varnothing$. For later use (in Section 12.5), we now record a lemma on the harmonic measure $\omega(z; \mathscr{A}_0)$ of $\mathscr{A}_0$ with respect to the unit disk. Roughly speaking, it is an estimate on the distance from $\mathscr{A}_0$ to $\mathscr{B}_0$, which increases as κ decreases.

LEMMA 7. For $z \in \mathscr{B}_0$, $\omega(z; \mathscr{A}_0) \leq 2\kappa$.

PROOF. By Lemma 6,

$$\omega(z; \mathscr{A}_0) \leq \omega(z; \mathscr{A}(2\varepsilon)) = \frac{\log |B(z)|}{\log(2\varepsilon)}.$$

But for $z \in \mathscr{B}_0$,

$$\log|B(z)| \geq \kappa \log \varepsilon \geq 2\kappa \log (2\varepsilon),$$

since $\varepsilon \leq \frac{1}{4}$. This gives the desired result.

Before turning directly to the proof of Carleson's lemma, it will be helpful to introduce some further notation. If $n \geq 1$ and $R_{nk}(i)$ [resp., R_{nk}] has the form

$$\{re^{i\theta} : a \leq r \leq b, \quad c \leq \theta \leq d\},$$

let $V_{nk}(i)$ [resp., V_{nk}] consist of the circular arc $\{r = a, c \leq \theta \leq d\}$ plus the two radial segments $\{a \leq r < 1, \theta = c, d\}$. Let $S_{nk}(i)$ [resp., S_{nk}] be the set

$$\{re^{i\theta} : a \leq r < 1, \quad c \leq \theta \leq d\}.$$

Finally, let $G_{nk}(i)$ be the union of $V_{nk}(i)$ and the boundaries of all the blocks $R_{ml}(j) \subset S_{nk}(i)$ such that $m < n + N$. This set $G_{nk}(i)$ will be known as the *grating* of $R_{nk}(i)$. (See Figure 4, where $N = 2$ for the sake of the illustration.)

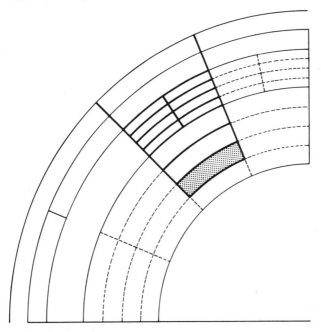

Figure 4   The grating $G_{12}(i)$ of a block $R_{12}(i)$ (shaded).

In terms of the set $\mathscr{A}_0$, we shall define a set E_{nk} *associated with* R_{nk}, as follows. A block $R_{ml}(j) \subset S_{nk} \cap \mathscr{A}_0$ will be called a *leading block* of S_{nk} if it is not contained in $S_{m'l'}(j')$ for any other block $R_{m'l'}(j') \subset S_{nk} \cap \mathscr{A}_0$. Then E_{nk} is defined as the union of V_{nk} and the gratings $G_{ml}(j)$ of all the leading blocks $R_{ml}(j)$ of S_{nk}. There can be at most a finite number of gratings in E_{nk}, since $\mathscr{B}_0$ contains some annular region $r_0 < |z| \leq 1$. Conceivably, E_{nk} may consist of V_{nk} alone.

Within each set S_{nk} we now construct a set F_{nk} called its *final set*. A rectangle $R_{ml} \subset S_{nk}$ will be called a *leading rectangle* of S_{nk} if R_{ml} meets $\mathcal{B}_0$ but is not contained in $S_{m'l'}$ for any other rectangle $R_{m'l'} \subset S_{nk}$ which meets $\mathcal{B}_0$. Thus the leading rectangles of S_{nk} are defined in terms of $\mathcal{B}_0$ just as the leading blocks are defined in terms of $\mathcal{A}_0$. Although S_{nk} may have no leading blocks, it does have finitely many leading rectangles, since $\mathcal{B}_0$ contains the annulus $r_0 < |z| \leq 1$. The *final set* F_{nk} of S_{nk} is now defined as the union of the sets E_{ml} associated with the leading rectangles R_{ml} of S_{nk}.

We are now equipped to describe a set Δ composed of certain radial segments and circular arcs in the open unit disk. The set Γ required for Carleson's lemma will be an appropriate subset of Δ. It will be shown in Section 12.5 that the arclength measure induced by Δ (hence that induced by Γ) satisfies the metrical condition (iii) of Carleson's lemma.

Let Δ_0 be the union of the boundaries of the blocks $R_{0k}(i)$, $k = 1, 2$; $i = 1$, ..., 2^{2N}. The sets S_{1k} ($k = 1, 2, 3, 4$) will be known as *residual sets of the zeroth generation*. Let Δ_1 be the union of the final sets F_{1k}, $k = 1, 2, 3, 4$. Adjacent to each grating $G_{nk}(i)$ in Δ_1, and within the region $S_{nk}(i)$ is a well determined set $S_{n+N, l}$, called a *residual set of the first generation*. Specifically, if

$$S_{nk}(i) = \{re^{i\theta} : a \leq r < 1, \quad c \leq \theta \leq c + 2^{-n-N}\pi\},$$

then the corresponding residual set is

$$S_{n+N, l} = \{re^{i\theta} : 1 - 2^{-n-N} \leq r < 1; \quad c \leq \theta \leq c + 2^{-n-N}\pi\}.$$

Now let Δ_2 be the union of the final sets $F_{n+N, l}$ of all the residual sets $S_{n+N, l}$ of the first generation. Adjacent to each grating in Δ_2 is a *residual set of the second generation*; let Δ_3 be the union of the final sets of these residual sets. The process continues until, after finitely many steps, a set Δ_M is constructed which contains no gratings. Let

$$\Delta = \Delta_0 \cup \Delta_1 \cup \cdots \cup \Delta_M.$$

It follows from the construction that $S_{nk} \cap \mathcal{A}_0 = \varnothing$ for each leading rectangle R_{nk} of a residual set of $(M-1)$st generation. Thus Δ surrounds $\mathcal{A}_0$ and separates it from the boundary of the disk.

We claim that Δ separates $\mathcal{A}_0$ from $\mathcal{B}_0$, in the sense that each continuous curve in $|z| < 1$ joining a point $z_1 \in \mathcal{A}_0$ to a point $z_2 \in \mathcal{B}_0$ must meet Δ (perhaps at z_1 or z_2). It is clear that z_1 and z_2 cannot belong to the same block $R_{nk}(i)$, since $\mathcal{A}_0 \cap \mathcal{B}_0 = \varnothing$. If $|z_2| \leq \frac{1}{2}$, then Δ_0 separates z_2 from z_1. If $|z_2| > \frac{1}{2}$, let S be the residual set of highest generation which contains z_2. Then $z_2 \in S_{nk}$ for some leading rectangle R_{nk} of S. If $z_1 \notin S_{nk}$, then $V_{nk} \subset \Delta$ separates z_1 from z_2. If $z_1 \in S_{nk}$, then $z_1 \in S_{ml}(j)$ for some leading block

$R_{ml}(j)$ of S_{nk}. But then the grating $G_{ml}(j)$ belongs to Δ and separates z_1 from z_2, since z_2 belongs to no residual set of generation higher than that of S.

Thus Δ separates $\mathscr{A}_0$ from $\mathscr{B}_0$. Now let $\hat{\Delta}$ be the set formed from Δ by deleting all interior points of $\mathscr{A}_0$ and of $\mathscr{B}_0$. It is clear that $\hat{\Delta}$ still separates $\mathscr{A}_0$ from $\mathscr{B}_0$. Let Ω be the union of all the components of the complement of $\hat{\Delta}$ which meet $\mathscr{A}_0$, and let Γ be the boundary of $\overline{\Omega}$. Then $\mathscr{A}(\varepsilon) \subset \mathscr{A}_0 \subset \Omega$; and $\mathscr{B}_0 \cap \Omega = \varnothing$, since $\hat{\Delta}$ separates $\mathscr{A}_0$ from $\mathscr{B}_0$. Consequently,

$$\varepsilon \le |B(z)| \le \varepsilon^\kappa \qquad \text{if} \quad z \in \Gamma.$$

Finally, let Γ have the orientation compatible with the counter-clockwise orientation of the blocks $R_{nk}(i)$. Then Γ is a contour having winding number 1 about each point in Ω, and in particular, about each zero of the Blaschke product $B(z)$. Hence Γ has properties (i) and (ii) of Carleson's lemma. We shall show in Section 12.5 that it also has property (iii).

12.5. ARCLENGTH OF Γ

To complete the proof of Carleson's lemma, it remains to show that the arclength measure μ induced by Γ is a Carleson measure. More specifically, it has to be shown that there is an absolute constant C such that

$$\mu(S) \le C\varepsilon^{-2}h$$

for each set S of the form

$$S = \{re^{i\theta} : 1 - h \le r < 1; \quad \theta_0 \le \theta \le \theta_0 + h\}.$$

We shall prove this (if κ is sufficiently small) for the arclength measure corresponding to the larger set Δ. In fact, the reader may find it convenient to identify μ with this larger measure.

It is clearly sufficient to show that

$$\mu(S_{nk}) \le C\varepsilon^{-2}2^{-n} \tag{9}$$

for all the sets S_{nk}. Observe that

$$\mu(R_{0k}) \le 2^N\pi < 16\pi\varepsilon^{-1} < 16\pi\varepsilon^{-2}, \qquad k = 1, 2,$$

since $\varepsilon/8 < 2^{-N+1}$. (Recall that N is the *smallest* integer with $2^{-N} \le \varepsilon/8$.) Hence we need only prove (9) for $n \ge 1$.

LEMMA 8. Let $G_{nk}(i)$ be an arbitrary grating, subtending an angle $\alpha = 2^{-n-N}\pi$ at the origin. Then

$$\mu(G_{nk}(i)) \le b\varepsilon^{-2}\alpha,$$

where b is an absolute constant.

PROOF. Directly from the definition, one sees that the radial segments in $G_{nk}(i)$ have total length less than $(N + 4)2^{-n-1}$, while the circular arcs have total length less than

$$(N + 1)2^{-n}\pi.$$

Since $N \geq 5$, these two estimates give

$$\mu(G_{nk}(i)) < 5N2^{-n} < b\varepsilon^{-2}\alpha,$$

where $b = 256$.

Next we consider rectangles which contain points of the set $\mathcal{B}_0$. The following lemma is the crucial step in the proof of (9). Recall that the parameter κ $(0 < \kappa < \frac{1}{2})$ is still at our disposal.

LEMMA 9. Suppose $\kappa \leq \kappa_0 = \frac{1}{8}[\cosh(\pi^2/2)]^{-2}$, let R_{nk} $(n \geq 1)$ be a rectangle which meets $\mathcal{B}_0$, and let the leading blocks of S_{nk} subtend angles $\alpha_1, \alpha_2, \ldots, \alpha_v$ at the origin. Then

$$\alpha_1 + \alpha_2 + \cdots + \alpha_v \leq \alpha/2,$$

where $\alpha = 2^{-n}\pi$ is the angle which R_{nk} subtends at the origin.

PROOF. We shall apply Lemma 5. Choose

$$z_0 = \rho^{1/3}e^{i\theta_0} \in R_{nk} \cap \mathcal{B}_0.$$

Let L_{nk} be the union of the leading blocks of S_{nk}, and let L_{nk}^* be the radial projection of L_{nk} onto the unit circle. Note that

$$L_{nk} \subset R = \{z : \rho < |z| < 1\},$$

since $(1 - 2^{-n-1})^3 < 1 - 2^{-n}$. Note also that L_{nk}^* has total length

$$|L_{nk}^*| = \alpha_1 + \alpha_2 + \cdots + \alpha_v.$$

Let $\omega_1(z)$ be the harmonic measure of L_{nk} with respect to R, and let $\omega_1^*(z)$ be the harmonic measure of L_{nk}^* with respect to R. Let $\omega(z; L_{nk})$ denote the harmonic measure of L_{nk} with respect to the unit disk. Since $L_{nk} \subset \mathcal{A}_0$, it follows from the maximum principle and Lemma 7 that

$$\omega_1(z_0) \leq \omega(z_0; L_{nk}) \leq \omega(z_0; \mathcal{A}_0) \leq 2\kappa. \tag{10}$$

On the other hand, the sector $|\theta - \theta_0| \leq \beta |\log \rho|$ contains S_{nk} (and hence L_{nk}) if $\beta |\log \rho| \geq 2^{-n}\pi$. Since

$$2^{-n} \leq |\log(1 - 2^{-n})| \leq |\log \rho| \leq 3|\log(1 - 2^{-n})| \leq 2^{1-n}\pi,$$

this will be the case if $\beta = \pi$. But $z_0 \notin L_{nk}$, since $\mathscr{A}_0 \cap \mathscr{B}_0 = \varnothing$, so Lemma 5 and (10) give

$$|L_{nk}^*| \leq |\log \rho| \cosh^2(\pi\beta/2)\omega_1(z_0)$$
$$< 2^{2-n}\pi\kappa \cosh^2(\pi^2/2) \leq \alpha/2$$

if $\kappa \leq \kappa_0$. This proves Lemma 9.

The inequality (9) can now be proved for an arbitrary residual set. As above, E^* will denote the radial projection of a set E onto the boundary of the unit disk. We assume henceforth that $\kappa \leq \kappa_0$.

LEMMA 10. Let S be a residual set of the mth generation, and let T be the union of the $(m+1)$st generation residual sets contained in S. Then

$$\mu(S \cap \Delta_{m+1}) \leq C\varepsilon^{-2}|S^*| \tag{11}$$

and

$$|T^*| \leq \tfrac{1}{2}|S^*|. \tag{12}$$

PROOF. Let R_{nk} be a leading rectangle of S, and let the leading blocks (if any) of S_{nk} subtend angles $\alpha_1, \alpha_2, \ldots, \alpha_\nu$ at the origin. The set E_{nk} associated with R_{nk} is the union of V_{nk} and the gratings of these leading blocks. Thus by Lemmas 8 and 9,

$$\mu(E_{nk}) \leq \mu(V_{nk}) + b\varepsilon^{-2}(\alpha_1 + \cdots + \alpha_\nu)$$
$$\leq 3\pi|E_{nk}^*| + \tfrac{1}{2}b\varepsilon^{-2}|E_{nk}^*|.$$

Summing over all the leading rectangles of S, we obtain (11). A similar application of Lemma 9 gives (12).

COROLLARY. There is an absolute constant C such that $\mu(S) \leq C\varepsilon^{-2}|S^*|$ for all residual sets S.

PROOF. Let S be a residual set of mth generation. Then successive applications of Lemma 10 show that

$$\mu(S \cap \Delta_{m+j}) \leq 2^{-j+1}C\varepsilon^{-2}|S^*|, \qquad j = 1, 2, \ldots, M - m.$$

The corollary then follows by addition over j.

It is now a simple matter to prove (9) for an arbitrary set S_{nk}. Indeed, let $\sigma_1, \sigma_2, \ldots, \sigma_p$ be the maximal residual sets contained in S_{nk}. That is, σ_j is a residual set in S_{nk}, not contained in any lower-generation residual set in S_{nk}. By the construction of Δ,

$$S_{nk} \cap \Delta \subset G_{nk} \cup \sigma_1 \cup \cdots \cup \sigma_p,$$

where G_{nk} is the union of the gratings $G_{nk}(i)$, $i = 1, \ldots, 2^{2N}$. Thus by Lemma 8 and the corollary to Lemma 10,

$$
\begin{aligned}
\mu(S_{nk}) &\leq \mu(G_{nk}) + \mu(\sigma_1) + \cdots + \mu(\sigma_p) \\
&\leq 2^N b\varepsilon^{-2} 2^{-n-N} \pi + C\varepsilon^{-2}(|\sigma_1{}^*| + \cdots + |\sigma_p{}^*|) \\
&\leq C_1 \varepsilon^{-2} |S_{nk}^*| .
\end{aligned}
$$

This establishes the required inequality (9), and the proof of Carleson's lemma is complete. This also finishes the proof of the corona theorem.

EXERCISES

1. Let A be the Banach algebra of functions $f(z)$ analytic in $|z| < 1$ and continuous in $|z| \leq 1$, with the uniform norm. Show that if $f_1, \ldots, f_n$ are functions in A with no common zero in $|z| \leq 1$, then there exist $g_1, \ldots, g_n$ in A such that

$$f_1(z)g_1(z) + \cdots + f_n(z)g_n(z) \equiv 1, \qquad |z| \leq 1.$$

2. Prove the following "zero–one" interpolation theorem. Let $\{a_n\}$ and $\{b_n\}$ be sequences of complex numbers in $|z| < 1$ such that $\sum (1 - |a_n|) < \infty$ and $\sum (1 - |b_n|) < \infty$, and let $A(z)$ and $B(z)$ denote the respective Blaschke products. Then there exists $f \in H^\infty$ such that $f(a_n) = 0$ and $f(b_n) = 1$ ($n = 1, 2, \ldots$) if and only if there is a number $\delta > 0$ such that

$$|A(z)| + |B(z)| \geq \delta, \qquad |z| < 1.$$

Under this condition, f can be chosen with $\|f\|_\infty \leq C$, a constant depending only on δ (Carleson [2]).

3. Use the zero–one interpolation theorem (Exercise 2) to prove the main interpolation theorem (Theorem 9.1) for H^∞: if $\{z_n\}$ is uniformly separated and $\{w_n\} \in \ell^\infty$, then there exists $f \in H^\infty$ such that $f(z_n) = w_n$, $n = 1, 2, \ldots$. (Carleson [2].)

NOTES

The corona theorem was conjectured by S. Kakutani as early as 1941. Carleson [2] gave a proof in 1961, basing part of the argument on unpublished work of D. J. Newman. In particular, the deduction of Theorem 12.1 from

Carleson's lemma (as presented in Section 12.2) is essentially due to Newman. Hörmander [2, 3] recently found a somewhat different approach which uses techniques borrowed from the theory of partial differential equations. Hall's lemma and some of its function-theoretic applications may be found in Hall [1]. The book of Gelfand, Raikov, and Shilov [1] is a good source for the basic theory of Banach algebras. The works of Wermer [1], Browder [1], and Gamelin [2] survey the theory of function algebras. Further information about the maximal ideal space of H^∞ is in the papers of Kakutani [1], Newman [2], Schark [1], Kerr-Lawson [1], and Hoffman [2].

The Rademacher functions $\varphi_1(t)$, $\varphi_2(t)$, ... are

$$\varphi_n(t) = \text{sgn}\{\sin(2^n\pi t)\}, \qquad 0 \le t \le 1.$$

Thus, for example,

$$\varphi_1(t) = \begin{cases} 1, & 0 < t < \tfrac{1}{2} \\ -1, & \tfrac{1}{2} < t < 1 \\ 0, & t = 0, \tfrac{1}{2}, 1. \end{cases}$$

In general, $\varphi_n(t)$ vanishes at all multiples of 2^{-n} and takes the values ± 1 elsewhere. Let R denote the set of all dyadic rationals in the interval $[0, 1]$; that is, numbers of the form $m2^{-n}$ ($m = 0, 1, \ldots, 2^n$; $n = 1, 2, \ldots$). The set R is countable, and so has measure zero.

Each number $t \in [0, 1] - R$ has a *unique* binary expansion

$$t = 0.b_1 b_2 b_3 \ldots, \qquad b_n = 0 \quad \text{or} \quad 1. \tag{1}$$

It is easy to see from the definition of the Rademacher functions that $\varphi_n(t) = 1$ if $b_n = 0$, while $\varphi_n(t) = -1$ if $b_n = 1$. The number t then determines a sequence of "signs" ± 1:

$$\varphi_1(t), \varphi_2(t), \varphi_3(t), \ldots; \tag{2}$$

and different t's generate different sign sequences. Since $t \notin R$, the sequence (2) cannot be eventually constant; it must assume each of the values $+1$ and -1 infinitely often. Furthermore, every sequence of signs $\{\varepsilon_n\}$ $(\varepsilon_n = \pm 1)$ not eventually constant is representable in the form (2) by a *unique* $t \in [0, 1] - R$. We have only to set $b_n = (1 - \varepsilon_n)/2$ and let the expansion (1) determine t. In short, the collection of all sign sequences $\{\varepsilon_n\}$ not eventually constant is in one–one correspondence with the set $[0, 1] - R$. It is now natural to define the measure of a given collection of sequences $\{\varepsilon_n\}$ as the Lebesgue measure of the corresponding subset of $[0, 1]$, provided this set is measurable. The set of all eventually constant sequences $\{\varepsilon_n\}$ is assigned measure zero. It becomes meaningful now to speak of "almost every sequence of signs."

The Rademacher functions form an orthonormal system over the interval $[0, 1]$:

$$\int_0^1 \varphi_n(t)\varphi_m(t)\, dt = \delta_{nm}.$$

More generally, if $n_1 < n_2 < \cdots < n_k$,

$$\int_0^1 \varphi_{n_1}(t)\varphi_{n_2}(t) \cdots \varphi_{n_k}(t)\, dt = 0. \tag{3}$$

To see this, observe that on each interval $(j/2^{n_k-1}, (j+1)/2^{n_k-1})$, the product $\varphi_{n_1}(t) \cdots \varphi_{n_{k-1}}(t)$ is constant, while $\varphi_{n_k}(t)$ takes the values $+1$ and -1 equally often.

THEOREM A.1 (Rademacher). Let $a_1, a_2, \ldots$ be complex numbers such that $\sum |a_n|^2 < \infty$. Then the series

$$\sum_{n=1}^{\infty} a_n \varphi_n(t) \tag{4}$$

converges almost everywhere.

Equivalently, the theorem says that for any square–summable sequence $\{a_n\}$, the series $\sum \pm a_n$ converges for almost every choice of signs.

PROOF OF THEOREM. Let $s_n(t) = \sum_{k=1}^n a_k \varphi_k(t)$. By the Riesz–Fischer theorem, there is an L^2 function $\Phi(t)$ such that

$$\lim_{n \to \infty} \int_0^1 |\Phi(t) - s_n(t)|^2\, dt = 0.$$

In particular, $\Phi(t)$ is integrable and (by the Schwarz inequality)

$$\lim_{n \to \infty} \int_\alpha^\beta s_n(t)\, dt = \int_\alpha^\beta \Phi(t)\, dt, \qquad 0 \le \alpha < \beta \le 1. \tag{5}$$

For almost every $t \in [0, 1]$, the indefinite integral $\int_0^t \Phi(u) \, du$ has a derivative equal to $\Phi(t)$; let $t_0 \notin R$ be such a point. For each integer m, t_0 is contained in a unique interval (α_m, β_m) of the form $(j/2^m, (j+1)/2^m)$. On the interval (α_m, β_m), $\varphi_k(t)$ is constant if $k \le m$, but takes the values ± 1 equally often if $k > m$. Thus

$$\int_{\alpha_m}^{\beta_m} [s_n(t) - s_m(t)] \, dt = 0 \qquad \text{for} \quad n > m.$$

In view of (5), we conclude that

$$0 = \lim_{n \to \infty} \int_{\alpha_m}^{\beta_m} [s_n(t) - s_m(t)] \, dt = \int_{\alpha_m}^{\beta_m} [\Phi(t) - s_m(t)] \, dt.$$

Hence, because $s_m(t)$ is constant on (α_m, β_m),

$$s_m(t_0) = \frac{1}{\beta_m - \alpha_m} \int_{\alpha_m}^{\beta_m} \Phi(t) \, dt \to \Phi(t_0) \qquad \text{as} \quad m \to \infty.$$

Thus $\{s_n(t)\}$ converges almost everywhere.

COROLLARY. If $\{\psi_n(x)\}$ is an orthonormal system in $L^2[a, b]$, and if $\sum |a_n|^2 < \infty$, then for almost every choice of signs $\{\varepsilon_n\}$, the series

$$\sum_{n=1}^{\infty} \varepsilon_n a_n \psi_n(x)$$

converges almost everywhere.

PROOF. By the Lebesgue monotone convergence theorem, the series $\sum_{n=1}^{\infty} |a_n \psi_n(x)|^2$ converges almost everywhere. Thus, for almost every $x \in [a, b]$, the series

$$\sum_{n=1}^{\infty} a_n \varphi_n(t) \psi_n(x) \tag{6}$$

converges for almost every $t \in [0, 1]$. Let E be the set of all points (t, x) at which (6) converges, and let $\Gamma(t, x)$ be the characteristic function of E. We have just observed that for almost every x, $\Gamma(t, x) = 1$ for almost all t. Fubini's theorem therefore gives

$$(b - a) = \int_a^b \int_0^1 \Gamma(t, x) \, dt \, dx = \int_0^1 \int_a^b \Gamma(t, x) \, dx \, dt.$$

If the corollary were false, however, the right-hand side would be less than $(b - a)$.

THEOREM A.2 (Khinchin's inequality). Suppose as in Theorem A.1 that $\sum |a_n|^2 < \infty$, and let $\Phi(t)$ be the sum (4). Then $\Phi \in L^p[0, 1]$ for every $p < \infty$, and

$$\left\{ \int_0^1 |\Phi(t)|^p \, dt \right\}^{1/p} \leq \left(\frac{p}{2} + 1 \right)^{1/2} \left\{ \sum_{n=1}^\infty |a_n|^2 \right\}^{1/2}. \tag{7}$$

If $p = 2m$ is an even integer, the constant $(p/2 + 1)^{1/2}$ can be replaced by $m^{1/2}$.

PROOF. First let $p = 2m$, and let the coefficients a_j be *real*. Then

$$\int_0^1 [s_n(t)]^{2m} \, dt = \sum \frac{(2m)!}{v_1! v_2! \cdots v_n!} a_1^{v_1} a_2^{v_2} \cdots a_n^{v_n} \int_0^1 [\varphi_1(t)]^{v_1} \cdots [\varphi_n(t)]^{v_n} dt,$$

where the sum is extended over all systems of nonnegative integers v_j such that $v_1 + \cdots + v_n = 2m$. In view of the orthogonality property (3), however, all the terms vanish except those for which all v_j are *even* integers (zero included). Thus the sum is equal to

$$\sum_{k_1 + \cdots + k_n = m} \frac{(2m)!}{(2k_1)!(2k_2)! \cdots (2k_n)!} a_1^{2k_1} a_2^{2k_2} \cdots a_n^{2k_n}.$$

On the other hand,

$$\sum_{k_1 + \cdots + k_n = m} \frac{m!}{k_1! k_2! \cdots k_n!} a_1^{2k_1} a_2^{2k_2} \cdots a_n^{2k_n} = \left\{ \sum_{j=1}^n a_j^2 \right\}^m.$$

But the ratio of the respective coefficients is

$$\frac{(2m)! \, k_1! k_2! \cdots k_n!}{m!(2k_1)!(2k_2)! \cdots (2k_n)!} \leq \frac{(2m)!}{m! 2^m} \leq m^m.$$

[If $k > 0$, $k!/(2k)! \leq (k + 1)^{-k} \leq 2^{-k}$; the inequality involving m^m may be verified by induction.] Therefore,

$$\int_0^1 |s_n(t)|^{2m} \, dt \leq m^m \left\{ \sum_{j=1}^n |a_j|^2 \right\}^m \tag{8}$$

if the a_j are real.

In the complex case, let $a_j = \alpha_j + i\beta_j$ and let

$$s_n = u_n + iv_n = \sum \alpha_j \varphi_j + i \sum \beta_j \varphi_j.$$

By Minkowski's inequality,

$$\left\{ \int_0^1 |s_n(t)|^{2m} \, dt \right\}^{1/m} = \left\{ \int (u_n^2 + v_n^2)^m \right\}^{1/m} \leq \left\{ \int u_n^{2m} \right\}^{1/m} + \left\{ \int v_n^{2m} \right\}^{1/m}$$

$$\leq m \sum_{j=1}^n \alpha_j^2 + m \sum_{j=1}^n \beta_j^2 = m \sum_{j=1}^n |a_j|^2,$$

which is equivalent to (8).

If $2m - 2 < p < 2m$, it follows that

$$\left\{\int_0^1 |s_n(t)|^p \, dt\right\}^{1/p} \leq \left\{\int_0^1 |s_n(t)|^{2m} \, dt\right\}^{1/2m} \leq m^{1/2}\left\{\sum_{j=1}^n |a_j|^2\right\}^{1/2}$$

$$\leq (p/2 + 1)^{1/2}\left\{\sum_{j=1}^n |a_j|^2\right\}^{1/2}.$$

Letting $n \to \infty$ here and in (8), we obtain (7) and its sharpened form for the case $p = 2m$. The argument also shows that $s_n \to \Phi$ in the L^p mean for every $p < \infty$.

THEOREM A.3 (Khinchin-Kolmogorov). If $\sum |a_n|^2 = \infty$, then the series $\sum_{n=1}^\infty a_n \varphi_n(t)$ diverges almost everywhere.

PROOF. Suppose, in fact, that the partial sums $s_n(t)$ are *bounded* on a set E of measure $|E| > 0$. Then

$$|s_n(t) - s_m(t)| \leq C, \qquad 1 \leq m < n; \, t \in E,$$

where C is a constant. It follows that

$$C^2 |E| \geq \int_E \left|\sum_{k=m+1}^n a_k \varphi_k(t)\right|^2 dt$$

$$= |E| \sum_{k=m+1}^n |a_k|^2 + 2 \sum_{\substack{j, k=m+1 \\ j<k}}^n a_j \overline{a_k} \int_E \varphi_j(t)\varphi_k(t) \, dt. \tag{9}$$

On the other hand, according to (3), the doubly indexed system of functions $\varphi_j(t)\varphi_k(t)$ $(1 \leq j < k < \infty)$ is orthonormal over $[0, 1]$. We may regard $\int_E \varphi_j(t)\varphi_k(t) \, dt$ as the "Fourier coefficient" of $\Gamma(t)$, the characteristic function of the set E. By Bessel's inequality, then,

$$\sum_{\substack{j,k=1 \\ j<k}}^\infty \left\{\int_E \varphi_j(t)\varphi_k(t) \, dt\right\}^2 \leq \int_0^1 [\Gamma(t)]^2 \, dt < \infty.$$

Therefore, for m sufficiently large,

$$\sum_{\substack{j, k=m+1 \\ j<k}}^\infty \left\{\int_E \varphi_j(t)\varphi_k(t) \, dt\right\}^2 < \left[\frac{|E|}{4}\right]^2. \tag{10}$$

It follows from (10) and the Schwarz inequality that the absolute value of the last term in (9) is no greater than

$$\frac{|E|}{2}\left\{\sum_{\substack{j, k=m+1 \\ j<k}}^n |a_j|^2|a_k|^2\right\}^{1/2} \leq \frac{|E|}{2} \sum_{k=m+1}^n |a_k|^2.$$

Introducing this estimate into (9), we find

$$\frac{|E|}{2} \sum_{k=m+1}^{n} |a_k|^2 \le C^2 |E|,$$

so that $\sum_{k=1}^{\infty} |a_k|^2 < \infty$.

The next theorem serves a special purpose in Section 4.6 and has other applications.

THEOREM A.4. Let $g_1(z)$, $g_2(z)$, ... be complex-valued functions each of which is continuous in $|z| \le 1$ except perhaps at a finite number of points on $|z| = 1$. Suppose

(i) $\sum_{n=1}^{\infty} |g_n(z)| < \infty$ in $|z| < 1$, the convergence being uniform in each disk $|z| \le \rho < 1$; and
(ii) for each N, $\sum_{n=N}^{\infty} |g_n(re^{i\theta})|^2 \to \infty$ uniformly in θ as $r \to 1$.

Then, for almost every choice of signs $\{\varepsilon_n\}$, the function

$$G(z) = \sum_{n=1}^{\infty} \varepsilon_n g_n(z)$$

has a radial limit almost nowhere. (That is, the θ's for which $\lim_{r \to 1} G(re^{i\theta})$ exists constitute a set of measure zero.)

PROOF. Because of hypothesis (i), the function

$$F(r, \theta, t) = \sum_{n=1}^{\infty} \varphi_n(t) g_n(re^{i\theta})$$

is well defined for all $r \in [0, 1)$, $\theta \in [0, 2\pi]$, and $t \in [0, 1]$. It is continuous in r and θ. Let

$$\mathscr{E} = \{(\theta, t) : \lim_{r \to 1} F(r, \theta, t) \quad \text{does not exist.}\}$$

We are going to show that for every $\theta \in [0, 2\pi]$, the "θ-section"

$$\mathscr{E}_\theta = \{t : (\theta, t) \in \mathscr{E}\}$$

has measure $|\mathscr{E}_\theta| = 1$. It will then follow from Fubini's theorem, as in the proof of the Corollary to Theorem A.1, that for almost every $t \in [0, 1]$, the set

$$\mathscr{E}^t = \{\theta : (\theta, t) \in \mathscr{E}\}$$

has measure 2π, as the theorem asserts.

Suppose, then, that $|\mathscr{E}_\theta| < 1$ for some fixed θ. Then the complementary set $\tilde{\mathscr{E}}_\theta$ has measure

$$\alpha = |\tilde{\mathscr{E}}_\theta| = 1 - |\mathscr{E}_\theta| > 0,$$

and $F(r, \theta, t)$ has a (finite) radial limit for all $t \in \tilde{\mathscr{E}}_\theta$. Because of the continuity of the functions $g_n(z)$, we may conclude that

$$F_N(r, \theta, t) = \sum_{n=N}^{\infty} \varphi_n(t) g_n(re^{i\theta}) \qquad (N = 1, 2, \ldots)$$

has a radial limit for all t in a set A of measure α, obtained from $\tilde{\mathscr{E}}_\theta$ by the deletion of a countable set.

Now, for $K = 1, 2, \ldots$, let

$$B_K = \{t \in A : |F(r, \theta, t)| \le K \quad \text{for all} \quad r < 1\}.$$

Then $B_1 \subset B_2 \subset \cdots$ and $\bigcup_{K=1}^{\infty} B_K = A$, so that $|B_K| \to |A| = \alpha$. Choose $K = K_1$ so large that $|B_{K_1}| \ge \frac{3}{4}\alpha$, and let $A_1 = B_{K_1}$. Then $|F(r, \theta, t)| \le K_1$ for all $t \in A_1$ and $r < 1$. Similarly, there are a set $A_2 \subset A_1$ and a constant K_2 such that $|A_2| \ge \frac{5}{8}\alpha$ and $|F_2(r, \theta, t)| \le K_2$ if $t \in A_2$ and $r < 1$. Proceeding inductively, we may construct a sequence of sets

$$A \supset A_1 \supset A_2 \supset \cdots \supset A_N \supset \cdots$$

and a sequence of constants $\{K_N\}$ such that

$$|A_N| \ge (\alpha/2)(1 + 2^{-N})$$

and

$$|F_N(r, \theta, t)| \le K_N, \qquad t \in A_N; \quad r < 1.$$

Finally, let

$$B = \bigcap_{N=1}^{\infty} A_N.$$

Then $|B| \ge \alpha/2 > 0$, and

$$\int_B |F_N(r, \theta, t)|^2 \, dt \le |B| K_N^2, \qquad r < 1; \quad N = 1, 2, \ldots. \tag{11}$$

We are now in a position to imitate the proof of Theorem A.3. First fix N so large that

$$\sum_{\substack{m, n = N \\ m < n}}^{\infty} \left\{ \int_B \varphi_m(t) \varphi_n(t) \, dt \right\}^2 < \frac{|B|^2}{16}.$$

In view of (11), an argument entirely similar to the proof of Theorem A.3 now gives

$$\frac{|B|}{2} \sum_{n=N}^{\infty} |g_n(re^{i\theta})|^2 \leq |B| K_N^2,$$

or

$$\sum_{n=N}^{\infty} |g_n(re^{i\theta})|^2 \leq 2K_N^2, \qquad r < 1.$$

But this contradicts hypothesis (ii). Hence $|\mathscr{E}_\theta| = 1$ for every θ, and the proof is complete.

THEOREM A.5. Let $a_0, a_1, \ldots$ be complex numbers such that

$$\limsup_{n \to \infty} |a_n|^{1/n} = 1.$$

(i) If $\sum |a_n|^2 < \infty$, then for almost every choice of signs $\{\varepsilon_n\}$,

$$f(z) = \sum_{n=0}^{\infty} \varepsilon_n a_n z^n \in H^p \qquad \text{for all } p < \infty.$$

(ii) If $\sum |a_n|^2 = \infty$, then for almost every choice of signs $\{\varepsilon_n\}$, $f(z)$ has a radial limit almost nowhere.

PROOF. Part (ii) is an immediate corollary of Theorem A.4, with $g_n(z) = a_n z^n$. We have only to observe that

$$\sum_{n=N}^{\infty} |a_n z^n|^2 = \sum_{n=N}^{\infty} |a_n|^2 r^{2n} \to \sum_{n=N}^{\infty} |a_n|^2 = \infty$$

as $r \to 1$.

(i) If $\sum |a_n|^2 < \infty$, then

$$g(z, t) = \sum_{n=1}^{\infty} \varphi_n(t) a_n z^n \in H^2$$

for every t, so (by Theorem 2.11) it need only be shown that for almost every $t \in [0, 1]$, the radial limit

$$G(\theta, t) = \lim_{r \to 1} g(re^{i\theta}, t) \tag{12}$$

belongs to L^p, as a function of θ, for all $p < \infty$. By Khinchin's inequality (Theorem A.2),

$$\int_0^1 |g(re^{i\theta}, t)|^p \, dt \leq (p/2 + 1)^{p/2} \left\{ \sum_{n=1}^{\infty} |a_n|^2 r^{2n} \right\}^{p/2}$$

$$\leq (p/2 + 1)^{p/2} \left\{ \sum_{n=1}^{\infty} |a_n|^2 \right\}^{p/2}.$$

On the other hand, Theorem A.1 implies that for each fixed θ, the limit (12) exists for almost every t. By Fatou's lemma, then,

$$\int_0^1 |G(\theta, t)|^p \, dt \le (p/2 + 1)^{p/2} \left\{ \sum_{n=1}^{\infty} |a_n|^2 \right\}^{p/2}, \qquad 0 \le \theta \le 2\pi.$$

Thus $|G(\theta, t)|^p$ is integrable over the rectangle, and by Fubini's theorem, $G(\theta, t) \in L^p$ for almost every t. To complete the proof, choose a sequence $p_1, p_2, \ldots \to \infty$, and let

$$E_k = \{t : G(\theta, t) \notin L^{p_k}\};$$

$$E = \{t : G(\theta, t) \notin L^p \quad \text{for some} \quad p < \infty\}.$$

Then $|E_k| = 0$ for each k, so that $E = \bigcup_{k=1}^{\infty} E_k$ is also of measure zero.

Even if $\sum |a_n|^2 \log n < \infty$, it can happen that $\sum \varepsilon_n a_n z^n \notin H^\infty$ for *every* choice of signs. However, the slightly stronger condition

$$\sum |a_n|^2 (\log n)^{1+\delta} < \infty \qquad (\delta > 0)$$

forces $\sum \varepsilon_n a_n z^n$ to be continuous in $|z| \le 1$ for almost every choice of signs. For the proofs we must refer the reader to the literature (see *Notes*).

NOTES

More information on Rademacher functions and related questions can be found in the books of Alexits [1], Kaczmarz and Steinhaus [1], and Zygmund [4]. Paley and Zygmund [1, 2] used Rademacher functions to prove various theorems in function theory. Proofs of the assertions in the last paragraph (above) may be found in Paley and Zygmund [1]. Theorem A.5 is due to Littlewood [4].

The Hardy–Littlewood "maximal theorem" becomes clearer if it is stated first in discrete form. Let $a_1, a_2, \ldots, a_n$ be given nonnegative numbers, and let $b_1, b_2, \ldots, b_n$ be the same numbers rearranged in nonincreasing order. For each fixed k $(k = 1, \ldots, n)$, let

$$\alpha_k = \max_{1 \le j \le k} \frac{1}{k - j + 1} \sum_{i=j}^{k} a_i$$

be the optimal average of successive a_i terminating with a_k; and let

$$\beta_k = \frac{1}{k} \sum_{i=1}^{k} b_i$$

be the corresponding quantities for the b_i. Let $s(x)$ be any nondecreasing function defined for all $x \ge 0$. Then, as Hardy and Littlewood showed,

$$\sum_{k=1}^{n} s(\alpha_k) \le \sum_{k=1}^{n} s(\beta_k).$$

Hardy and Littlewood interpreted the a_i as cricket scores and s as the batsman's "satisfaction function." The theorem then says that the batsman's "total satisfaction" is maximized if he plays a given collection of innings in decreasing order.

In order to state a continuous version of this theorem, it is necessary to define the "rearrangement" of a function. Let $f(x)$ be nonnegative and integrable over a finite interval $[0, a]$, and let $\mu(y)$ be the measure of the set in which $f(x) > y$. Note that $\mu(y)$ is nonincreasing. Two functions $f_1(x)$ and $f_2(x)$ are said to be *equimeasurable* if they give rise to the same function $\mu(y)$. It is then clear from the definition of Lebesgue integral that

$$\int_0^a f_1(x) \, dx = \int_0^a f_2(x) \, dx.$$

If $\mu(y)$ is associated as above with $f(x)$, its inverse function $f^*(x) = \mu^{-1}(x)$, normalized so that $f^*(x) = f^*(x+)$, is called the *decreasing rearrangement* of $f(x)$. It is easy to see that $f^*(x)$ and $f(x)$ are equimeasurable. Finally, let

$$A(x, \xi) = A(x, \xi; f) = \frac{1}{x - \xi} \int_\xi^x f(t) \, dt, \qquad 0 \le \xi < x;$$

and let

$$\Theta(x) = \Theta(x; f) = \sup_{0 \le \xi < x} A(x, \xi; f).$$

The maximal theorem may now be stated as follows.

THEOREM B.1 (Hardy-Littlewood maximal theorem). If $s(y)$ is any nondecreasing function defined for $y \ge 0$, then

$$\int_0^a s(\Theta(x; f)) \, dx \le \int_0^a s(\Theta(x; f^*)) \, dx.$$

Hardy and Littlewood used a limiting process to base a proof on the discrete form of the theorem. Shortly afterwards, F. Riesz showed that in fact

$$\Theta^*(x; f) \le \Theta(x; f^*), \qquad 0 < x \le a, \tag{1}$$

which immediately implies the Hardy–Littlewood result. Riesz's proof of (1) makes use of his "rising sun lemma," which may be stated as follows.

LEMMA. Let $g(x)$ be continuous on the interval $[0, a]$, and let E be the set of all x in $(0, a)$ for which there exists ξ in $[0, x)$ with $g(\xi) < g(x)$. Then E is an open set: $E = \bigcup(a_k, b_k)$, where the intervals (a_k, b_k) are disjoint; and $g(a_k) \le g(b_k)$.

PROOF OF LEMMA. Since the inequalities $\xi < x$ and $g(\xi) < g(x)$ are not disturbed by small changes in x, E is open. To show that $g(a_k) \le g(b_k)$, choose $x_0 \in (a_k, b_k)$ and consider the point x_1 where $g(x)$ attains its minimum in $[0, x_0]$. Then $x_1 \notin (a_k, x_0]$, since these points are in E. Thus $x_1 \in [0, a_k]$. But since $a_k \notin E$, $g(a_k) \leq g(x)$ for all $x \in [0, a_k)$. Hence $x_1 = a_k$, and in particular, $g(a_k) \leq g(x_0)$. Now let $x_0 \to b_k$ to conclude that $g(a_k) \leq g(b_k)$.

PROOF OF (1). For fixed $y_0 \geq 0$, apply the lemma to the function

$$g(x) = \int_0^x f(t)\, dt - y_0 x.$$

E is then the set of all x for which there exists ξ in $[0, x)$ with $A(x, \xi; f) > y_0$. In other words,

$$E = \{x : \Theta(x;f) > y_0\}. \tag{2}$$

On the other hand, the condition $g(a_k) \leq g(b_k)$ says that $A(b_k, a_k; f) \geq y_0$. Thus

$$\int_E f(x)\, dx = \sum_k \int_{a_k}^{b_k} f(x)\, dx \geq y_0 \sum_k (b_k - a_k) = y_0\, m(E), \tag{3}$$

where $m(E)$ denotes the measure of E. Now let

$$f_1(x) = \begin{cases} f(x), & x \in E \\ 0, & x \notin E. \end{cases}$$

Since $f_1(x) \leq f(x)$, it follows that $f_1^*(x) \leq f^*(x)$. Thus by (3),

$$\Theta(m(E); f^*) = A(m(E), 0; f^*) \geq A(m(E), 0; f_1^*)$$

$$= \frac{1}{m(E)} \int_0^a f_1(x)\, dx = \frac{1}{m(E)} \int_E f(x)\, dx \geq y_0. \tag{4}$$

All this is for arbitrary $y_0 \geq 0$. Given $x_0 \in (0, a]$, choose $y_0 = \Theta^*(x_0; f)$. Then by (2) and the definition of Θ^*, $m(E) = x_0$. Hence (4) gives

$$\Theta(x_0; f^*) \geq y_0 = \Theta^*(x_0; f),$$

which proves (1).

The following application of the maximal theorem will serve to illustrate its usefulness.

THEOREM B.2. If $f(x)$ belongs to L^p $(1 < p < \infty)$ over some interval $[0, a]$, then the maximal function $\Theta(x) = \Theta(x; |f|)$ is also in L^p, and

$$\|\Theta\|_p \leq \frac{p}{p-1} \|f\|_p.$$

PROOF. Apply the maximal theorem with $s(y) = y^p$, note that $\Theta(x; f^*) = A(x, 0; f^*)$, and use the following inequality.

LEMMA (Hardy's inequality). If $1 < p < \infty$, $g(x)$ is in L^p over $(0, \infty)$, and

$$G(x) = \frac{1}{x} \int_0^x g(t)\, dt,$$

then $G \in L^p$ and

$$\|G\|_p \le \frac{p}{p-1} \|g\|_p.$$

PROOF OF LEMMA. Fix $a > 0$. Since

$$G(x) = \frac{1}{a} \int_0^a g\left(\frac{xt}{a}\right) dt,$$

the continuous form of Minkowski's inequality gives

$$\left\{ \int_0^a |G(x)|^p\, dx \right\}^{1/p} \le \frac{1}{a} \int_0^a \left\{ \int_0^a \left| g\left(\frac{xt}{a}\right) \right|^p dx \right\}^{1/p} dt$$

$$\le a^{1/p-1} \int_0^a t^{-1/p}\, dt \left\{ \int_0^a |g(u)|^p\, du \right\}^{1/p}$$

$$= \frac{p}{p-1} \left\{ \int_0^a |g(u)|^p\, du \right\}^{1/p}.$$

Now let $a \to \infty$ to obtain Hardy's inequality.

The next theorem is essentially a restatement of Theorem B.2 in a form convenient for certain applications (see Section 1.6).

THEOREM B.3. Let $f(x)$ be periodic with period 2π, and suppose $f \in L^p = L^p(0, 2\pi)$, $1 < p < \infty$. Then

$$F(x) = \sup_{0 < |T| \le \pi} \left| \frac{1}{T} \int_0^T f(x + t)\, dt \right|$$

is also in L^p, and $\|F\|_p \le C_p \|f\|_p$, where C_p is a constant depending only on p.

PROOF. It is clearly enough to assume $f(x) \ge 0$ and to consider

$$F_1(x) = \sup_{0 < T \le \pi} \frac{1}{T} \int_{-T}^0 f(x + t)\, dt$$

$$= \sup_{x - \pi \le \xi < x} \frac{1}{x - \xi} \int_\xi^x f(t)\, dt.$$

This function is easily compared with the maximal function $\Theta(x) = \Theta(x;f)$ associated with f. For $x \geq \pi$, it follows from the definitions that $F_1(x) \leq \Theta(x)$. For $0 \leq x < \pi$, the periodicity of f may be used to show that $F_1(x) \leq \Theta(x+2\pi)$. Hence by Theorem B.2,

$$\int_0^{2\pi} |F_1(x)|^p \, dx \leq \int_\pi^{3\pi} |\Theta(x)|^p \, dx \leq \left(\frac{p}{p-1}\right)^p \int_0^{3\pi} |f(x)|^p \, dx,$$

as desired. The constant C_p can be taken as $4^{1/p}p(p-1)^{-1}$.

NOTES

Maximal theorems are the invention of Hardy and Littlewood [4], who proved Theorem B.1 and pointed out its applications to function theory. The simple proof given above is due to F. Riesz [8]. See also Hardy, Littlewood, and Pólya [1]. Riesz had introduced the "rising sun lemma" as a tool in differentiation theory (see, for example, Boas [2]). Hardy's inequality and related results are in Hardy, Littlewood, and Pólya [1].

REFERENCES

Ahlfors, L. V.
  [1] Bounded analytic functions. *Duke Math. J.* **14** (1947), 1–11.
  [2] "Complex Analysis," 2nd ed. McGraw-Hill, New York, 1966.

Ahlfors, L. V., and Sario, L.
  [1] "Riemann Surfaces." Princeton Univ. Press, Princeton, New Jersey, 1960.

Ahlfors, L. V., and Weill, G.
  [1] A uniqueness theorem for Beltrami equations. *Proc. Amer. Math. Soc.* **13** (1962), 975–978.

Akutowicz, E. J.
  [1] A qualitative characterization of Blaschke products in a half-plane. *Amer. J. Math.* **78** (1956), 677–684.
  [2] Schwarz's lemma in the Hardy class H^1. *Rend. Circ. Mat. Palermo* **8** (1959), 185–191.

Akutowicz, E. J., and Carleson, L.
  [1] The analytic continuation of interpolatory functions. *J. Analyse Math.* **7** (1959–60), 223–247.

Alexits, G.
  [1] "Convergence Problems of Orthogonal Series." Pergamon Press, Oxford, 1961.

Banach, S.
  [1] "Théorie des opérations linéaires" (Warsaw, 1932). Reprinted by Chelsea, New York, 1955.

Bary, N. K.
[1] "A Treatise on Trigonometric Series" (Moscow, 1961). English transl.: Pergamon Press, New York, 1964.

Beurling, A.
[1] On two problems concerning linear transformations in Hilbert space. *Acta Math.* **81** (1949), 239–255.

Blaschke, W.
[1] Eine Erweiterung des Satzes von Vitali über Folgen analytischer Funktionen. *S.-B. Sächs. Akad. Wiss. Leipzig Math.-Natur. Kl.* **67** (1915), 194–200.

Boas, R. P.
[1] Isomorphism between H^p and L^p. *Amer. J. Math.* **77** (1955), 655–656.
[2] "A Primer of Real Functions." Carus Math. Monograph No. 13, Math. Assoc. of Amer., 1960.
[3] Majorant problems for trigonometric series. *J. Analyse Math.* **10** (1962–63), 253–271.

Bonsall, F. F.
[1] Dual extremum problems in the theory of functions. *J. London Math. Soc.* **31** (1956), 105–110.

Browder, A.
[1] "Introduction to Function Algebras." Benjamin, New York, 1969.

Calderón, A. P.
[1] On theorems of M. Riesz and Zygmund. *Proc. Amer. Math. Soc.* **1** (1950), 533–535.

Cantor, D. G.
[1] Power series with integral coefficients. *Bull. Amer. Math. Soc.* **69** (1963), 362–366.

Carathéodory, C., and Fejér, L.
[1] Remarque sur le théorème de M. Jensen. *C. R. Acad. Sci. Paris* **145** (1907), 163–165.
[2] Über den Zusammenhang der Extremen von harmonischen Funktionen mit ihren Koeffizienten und über den Picard–Landauschen Satz. *Rend. Circ. Mat. Palermo* **32** (1911), 218–239.

Cargo, G. T.
[1] The boundary behavior of Blaschke products. *J. Math. Anal. Appl.* **5** (1962), 1–16.
[2] Angular and tangential limits of Blaschke products and their successive derivatives. *Canad. J. Math.* **14** (1962), 334–348.

Carleson, L.
[1] An interpolation problem for bounded analytic functions. *Amer. J. Math.* **80** (1958), 921–930.
[2] Interpolations by bounded analytic functions and the corona problem. *Ann. of Math.* **76** (1962), 547–559.
[3] Interpolations by bounded analytic functions and the corona problem. *Proc. Internat. Congr. Math. Stockholm*, 1962, pp. 314–316.

Caughran, J. G.
[1] Nonuniqueness of extremal kernels. *Proc. Amer. Math. Soc.* **18** (1967), 957–958.
[2] Factorization of analytic functions with H^p derivative. *Duke Math. J.* **36** (1969), 153–158.

Caveny, J.
[1] Bounded Hadamard products of H^p functions. *Duke Math. J.* **33** (1966), 389–394.
[2] Absolute convergence factors for H^p series. *Canad. J. Math.* **21** (1969), 187–195.

Coifman, R., and Weiss, G.
[1] A kernel associated with certain multiply connected domains and its applications to factorization theorems. *Studia Math.* **28** (1966), 31–68.

Collingwood, E. F., and Lohwater, A. J.
[1] "The Theory of Cluster Sets." Cambridge Univ. Press, London and New York, 1966.

Day, M. M.
[1] The spaces L^p with $0 < p < 1$. *Bull. Amer. Math. Soc.* **46** (1940), 816–823.

Doob, J. L.
[1] A minimum problem in the theory of analytic functions. *Duke Math. J.* **8** (1941), 413–424.

Douglas, R. G., and Rudin, W.
[1] Approximation by inner functions. *Pacific J. Math.* **31** (1969), 313–320.

Dunford, N., and Schwartz, J. T.
[1] "Linear Operators, Part I." Interscience, New York, 1958.

Duren, P. L.
[1] Smoothness of functions generated by Riesz products. *Proc. Amer. Math. Soc.* **16** (1965), 1263–1268.
[2] Extension of a theorem of Carleson. *Bull. Amer. Math. Soc.* **75** (1969), 143–146.
[3] On the multipliers of H^p spaces. *Proc. Amer. Math. Soc.* **22** (1969), 24–27.
[4] On the Bloch–Nevanlinna conjecture. *Colloq. Math.* **20** (1969), 295–297.
[5] Schwarzian derivatives and mappings onto Jordan domains. *Ann. Acad. Sci. Fenn.* (To be published.)

Duren, P. L., Romberg, B. W., and Shields, A. L.
[1] Linear functionals on H^p spaces with $0 < p < 1$. *J. Reine Angew. Math.* **238** (1969), 32–60.

Duren, P. L., Shapiro, H. S., and Shields, A. L.
[1] Singular measures and domains not of Smirnov type. *Duke Math. J.* **33** (1966), 247–254.

Duren, P. L., and Shields, A. L.
[1] Properties of H^p $(0 < p < 1)$ and its containing Banach space. *Trans. Amer. Math. Soc.* **141** (1969), 255–262.
[2] Coefficient multipliers of H^p and B^p spaces. *Pacific J. Math.* **32** (1970), 69–78.

Duren, P. L., and Taylor, G. D.
[1] Mean growth and coefficients of H^p functions. *Illinois J. Math.* **14** (1970).

Earle, C. J., and Marden, A.
[1] Projections to automorphic functions. *Proc. Amer. Math. Soc.* **19** (1968), 274–278.
[2] On Poincaré series with application to H^p spaces on bordered Riemann surfaces. *Illinois J. Math.* **13** (1969), 202–219.

Egerváry, E.
[1] Über gewisse Extremumprobleme der Funktionentheorie. *Math. Ann.* **99** (1928), 542–561.

Epstein, B., and Minker, J.
[1] Extremal interpolatory problems in the unit disc. *Proc. Amer. Math. Soc.* **11** (1960), 777–784.

Evgrafov, M. A.
[1] Behavior of power series for functions of class H_δ on the boundary of the circle of convergence. *Izv. Akad. Nauk SSSR Ser. Mat.* **16** (1952), 481–492 (in Russian).

Fatou, P.
[1] Séries trigonométriques et séries de Taylor. *Acta Math.* **30** (1906), 335–400.

Fejér, L.
[1] Über gewisse Minimumprobleme der Funktionentheorie. *Math. Ann.* **97** (1926), 104–123.

Fejér, L., and Riesz, F.
[1] Über einige funktionentheoretische Ungleichungen. *Math. Z.* **11** (1921), 305–314.

Fichtenholz, G.
[1] Sur l'intégrale de Poisson et quelques questions qui s'y rattachent. *Fund. Math.* **13** (1929), 1–33.

Fisher, S. D.
[1] Rational functions H^∞ and H^p on infinitely connected domains. *Illinois J. Math.* **12** (1968), 513–523.

Flett, T. M.
[1] Mean values of power series. *Pacific J. Math.* **25** (1968), 463–494.
[2] On the rate of growth of mean values of holomorphic and harmonic functions. *Proc. London Math. Soc.* **20** (1970).

Forelli, F.
[1] The Marcel Riesz theorem on conjugate functions. *Trans. Amer. Math. Soc.* **106** (1963), 369–390.
[2] The isometries of H^p. *Canad. J. Math.* **16** (1964), 721–728.
[3] Bounded holomorphic functions and projections. *Illinois J. Math.* **10** (1966), 367–380.
[4] Extreme points in $H^1(R)$. *Canad. J. Math.* **19** (1967), 312–320.

Frostman, O.
[1] Potential d'équilibre et capacité des ensembles avec quelques applications à la théorie des fonctions. *Meddel. Lunds Univ. Mat. Sem.* **3** (1935), 1–118.
[2] Sur les produits de Blaschke. *Kungl. Fysiografiska Sällskapets i Lund Förhandlingar* **12** (1942), no. 15, 169–182.

Gabriel, R. M.
[1] Some results concerning the integrals of moduli of regular functions along curves of certain types. *Proc. London Math. Soc.* **28** (1928), 121–127.
[2] An inequality concerning the integrals of positive subharmonic functions along certain circles. *J. London Math. Soc.* **5** (1930), 129–131.

Gamelin, T. W.
[1] H^p spaces and extremal functions in H^1. *Trans. Amer. Math. Soc.* **124** (1966), 158–167.
[2] "Uniform Algebras." Prentice-Hall, Englewood Cliffs, New Jersey, 1969.

Gamelin, T. W., and Lumer, G.
[1] Theory of abstract Hardy spaces and the universal Hardy class. *Advances in Math.* **2** (1968), 118–174.

Gamelin, T. W., and Voichick, M.
[1] Extreme points in spaces of analytic functions. *Canad. J. Math.* **20** (1968), 919–928.

Garabedian, P. R.
[1] Schwarz's lemma and the Szegö kernel function. *Trans. Amer. Math. Soc.* **67** (1949), 1–35.

Gaudry, G. I.
[1] H^p multipliers and inequalities of Hardy and Littlewood. *J. Austral. Math. Soc.* **10** (1969), 23–32.

Gelfand, I. M., Raikov, D., and Shilov, G.
[1] "Commutative Normed Rings" (Moscow, 1960). English transl.: Chelsea, New York, 1964.

Geronimus, Ja. L.
[1] Sur le problème des coefficients pour les fonctions bornées. *Dokl. Akad. Nauk SSSR* **14** (1937), 97–98.
[2] On some extremal problems. *Izv. Akad. Nauk SSSR Ser. Mat.* **1** (1937), 185–202 (in Russian).
[3] On a problem of F. Riesz and the generalized problem of Chebyshev–Korkin–Zolotarev. *Izv. Akad. Nauk SSSR Ser. Mat.* **3** (1939), 279–288 (in Russian).

Gleason, A. M., and Whitney, H.
[1] The extension of linear functionals defined on H^∞. *Pacific J. Math.* **12** (1962), 163–182.

Goffman, C., and Pedrick, G.
[1] "First Course in Functional Analysis." Prentice-Hall, Englewood Cliffs, New Jersey, 1965.

Goldberg, R. R.
[1] "Fourier Transforms." Cambridge Univ. Press, London and New York, 1961.

Golubev, V.
[1] Sur la correspondance des frontières dans la représentation conforme. *Mat. Sb.* **32** (1924), 55–57 (in Russian).

Goluzin, G. M.
[1] On the problem of Carathéodory–Fejér and similar problems. *Mat. Sb.* **18** (60) (1946), 213–226 (in Russian).
[2] Estimates for analytic functions with bounded mean modulus. *Trudy Mat. Inst. Steklov* **18** (1946), 1–87 (in Russian).
[3] "Geometric Theory of Functions of a Complex Variable" (Moscow, 1952). German transl.: Deutscher Verlag, Berlin, 1957. 2nd ed. (Moscow, 1966) English transl.: American Mathematical Society, Providence, Rhode Island, 1969.

Gronwall, T. H.
[1] On the maximum modulus of an analytic function. *Ann. of Math.* **16** (1915), 77–81.

Gwilliam, A. E.
[1] On Lipschitz conditions. *Proc. London Math. Soc.* **40** (1936), 353–364.

Hall, T.
[1] Sur la mesure harmonique de certains ensembles. *Arkiv. Mat. Astr. Fys.* **25A** (1937), No. 28, 8 pp.

Hardy, G. H.
[1] A theorem concerning Taylor's series. *Quart. J. Pure Appl. Math.* **44** (1913), 147–160.
[2] The mean value of the modulus of an analytic function. *Proc. London Math. Soc.* **14** (1915), 269–277.
[3] Remarks on three recent notes in the Journal. *J. London Math. Soc.* **3** (1928), 166–169.

Hardy, G. H., and Littlewood, J. E.
[1] Some new properties of Fourier constants. *Math. Ann.* **97** (1926), 159–209.
[2] Some properties of fractional integrals, I. *Math. Z.* **27** (1928), 565–606.
[3] A convergence criterion for Fourier series. *Math. Z.* **28** (1928), 612–634.
[4] A maximal theorem with function-theoretic applications. *Acta Math.* **54** (1930) 81–116.
[5] Some properties of fractional integrals, II. *Math. Z.* **34** (1932), 403–439.
[6] Some properties of conjugate functions. *J. Reine Angew. Math.* **167** (1931), 405–423.
[7] Notes on the theory of series (XX): Generalizations of a theorem of Paley. *Quart. J. Math., Oxford Ser.* **8** (1937), 161–171.
[8] Theorems concerning mean values of analytic or harmonic functions. *Quart. J. Math., Oxford Ser.* **12** (1941), 221–256.

Hardy, G. H., Littlewood, J. E., and Pólya, G.
[1] "Inequalities," 2nd ed. Cambridge Univ. Press, London and New York, 1952.

Havin, V. P.
[1] On analytic functions representable by an integral of Cauchy–Stieltjes type. *Vestnik Leningrad. Univ.* **13** (1958), no. 1, 66–79 (in Russian).
[2] Relations between certain classes of functions regular in the unit circle. *Vestnik Leningrad. Univ.* **17** (1962), no. 1, 102–110 (in Russian).
[3] Boundary properties of integrals of Cauchy type and of conjugate harmonic functions in regions with rectifiable boundary. *Mat. Sb.* **68** (110) (1965), 499–517 (in Russian).
[4] Spaces of analytic functions. *In* "Math. Analysis 1964," pp. 76–164. Akad. Nauk SSSR Inst. Naučn. Informacii, Moscow, 1966 (in Russian).

Havinson, S. Ja.
[1] On an extremal problem in the theory of analytic functions. *Uspehi Mat. Nauk* **4** (1949), no. 4 (32), 158–159 (in Russian).
[2] On some extremal problems in the theory of analytic functions. *Moskov. Gos. Univ. Učenye Zapiski* **148**, Matem. **4** (1951), 133–143 (in Russian). English transl.: *Amer. Math. Soc. Transl.* (2) **32** (1963), 139–154.
[3] Extremal problems for certain classes of analytic functions in finitely connected regions. *Mat. Sb.* **36** (78) (1955), 445–478 (in Russian). English transl.: *Amer. Math. Soc. Transl.* (2) **5** (1957), 1–33.
[4] Analytic functions of bounded type. *In* "Math. Analysis 1963," pp. 5–80. Akad. Nauk SSSR Inst. Naučn. Informacii, Moscow, 1965 (in Russian).

Hayman, W. K.
[1] "Multivalent Functions." Cambridge Univ. Press, London and New York, 1958.
[2] Interpolation by bounded functions. *Ann. Inst. Fourier (Grenoble)* **8** (1959), 277–290.
[3] On the characteristic of functions meromorphic in the unit disk and of their integrals. *Acta Math.* **112** (1964), 181–214.

Hedlund, J. H.
[1] Multipliers of H^p spaces. *J. Math. Mech.* **18** (1969), 1067–1074.
[2] Multipliers of H^1 and Hankel matrices. *Proc. Amer. Math. Soc.* **22** (1969), 20–23.

Heins, M.
  [1] On the theorem of Szegö–Solomentsev. *Math. Scand.* **20** (1967), 281–289.
  [2] "Hardy Classes on Riemann Surfaces." Lecture Notes in Mathematics, No. 98. Springer, Berlin, 1969.

Helson, H.
  [1] On a theorem of F. and M. Riesz. *Colloq. Math.* **3** (1955), 113–117.
  [2] Conjugate series and a theorem of Paley. *Pacific J. Math.* **8** (1958), 437–446.
  [3] "Lectures on Invariant Subspaces." Academic Press, New York, 1964.

Herglotz, G.
  [1] Über Potenzreihen mit positivem, reellen Teil im Einheitskreis. *S.-B. Sächs. Akad. Wiss. Leipzig Math.-Natur. Kl.* **63** (1911), 501–511.

Hille, E., and Tamarkin, J. D.
  [1] On a theorem of Paley and Wiener. *Ann. of Math.* **34** (1933), 606–614.
  [2] A remark on Fourier transforms and functions analytic in a half-plane. *Compositio Math.* **1** (1934), 98–102.
  [3] On the absolute integrability of Fourier transforms. *Fund. Math.* **25** (1935), 329–352.

Hoffman, K.
  [1] "Banach Spaces of Analytic Functions." Prentice-Hall, Englewood Cliffs, New Jersey, 1962.
  [2] Bounded analytic functions and Gleason parts. *Ann. of Math.* **86** (1967), 74–111.

Hörmander, L.
  [1] L^p estimates for (pluri-) subharmonic functions. *Math. Scand.* **20** (1967), 65–78.
  [2] Generators for some rings of analytic functions. *Bull. Amer. Math. Soc.* **73** (1967), 943–949.
  [3] The work of Carleson and Hoffman on bounded analytic functions in the disc. Lecture notes, Current Literature Seminar, Institute for Advanced Study, Princeton, 1967.

Kabaïla, V.
  [1] On interpolation by functions of class H_δ. *Uspehi Mat. Nauk* **13** (1958), no. 1 (79), 181–188 (in Russian).
  [2] Refinement of some theorems on interpolation in the class H_δ. *Vilniaus Univ. Darbai* **33**, *Mat. -Fiz.* **9** (1960), 15–19 (in Lithuanian; Russian summary).
  [3] Some interpolation problems in the class H_δ for $\delta < 1$. *In* "Issledovaniya po sovremennym problemam teorii funkciĭ kompleksnogo peremennogo," Moscow, 1961, pp. 180–187 (in Russian).
  [4] On the paper "Some interpolation problems in the class H_δ for $\delta < 1$." *Litovsk. Mat. Sb.* **2** (1962), no. 2, 145–148 (in Russian).
  [5] Interpolation sequences for the H_p classes in the case $p < 1$. *Litovsk. Mat. Sb.* **3** (1963), no. 1, 141–147 (in Russian).

Kaczmarz, S.
  [1] On some classes of Fourier series. *J. London Math. Soc.* **8** (1933), 39–46.

Kaczmarz, S., and Steinhaus, H.
  [1] "Theorie der Orthogonalreihen" (Warsaw, 1935). Reprinted by Chelsea, New York, 1951.

Kahane, J.-P.
  [1] Trois notes sur les ensembles parfaits linéaires. *Enseignement Math.* **15** (1969), 185–192.

Kakeya, S.
[1] General mean modulus of analytic function. *Proc. Phys.-Math. Soc. Japan* **3** (1921), 48–58.
[2] Maximum modulus of some expressions of limited analytic functions. *Trans. Amer. Math. Soc.* **22** (1921), 489–504.

Kakutani, S.
[1] Rings of analytic functions. *In* "Lectures on Functions of a Complex Variable," pp. 71–83. Univ. of Michigan Press, Ann Arbor, 1955.

Kas'yanyuk, S. A.
[1] On functions of class A and H_δ in a circular ring. *Mat. Sb.* **42** (84) (1957), 301–326 (in Russian).
[2] Letter to the editor. *Mat. Sb.* **47** (89) (1959), 141–142 (in Russian).

Kawata, T.
[1] On analytic functions regular in the half-plane (I; II). *Japan. J. Math.* **13** (1937), 421–430; 483–491.

Keldysh, M. V.
[1] On a class of extremal polynomials. *Dokl. Akad. Nauk SSSR* **4** (13) (1936), no. 4 (108), 163–166 (in Russian).

Keldysh, M. V., and Lavrentiev, M. A.
[1] Sur la représentation conforme des domaines limités par des courbes rectifiables. *Ann. Sci. École Norm. Sup.* **54** (1937), 1–38.

Kerr-Lawson, A.
[1] A filter description of the homomorphisms of H^∞. *Canad. J. Math.* **17** (1965), 734–757.
[2] Some lemmas on interpolating Blaschke products and a correction. *Canad. J. Math.* **21** (1969), 531–534.

Kolmogorov, A.
[1] Sur les fonctions harmoniques conjuguées et les séries de Fourier. *Fund. Math.* **7** (1925), 24–29.

Köthe, G.
[1] "Topologische Lineare Räume I." Springer, Berlin, 1960.

Krein, M., and Rechtman, P.
[1] On a problem of Nevanlinna–Pick. *Trudy Odessa Univ., Mat.* **2** (1938), 63–68 (in Ukrainian; French summary).

Krylov, V. I.
[1] On functions regular in a half-plane. *Mat. Sb.* **6** (48) (1939), 95–138. English transl.: *Amer. Math. Soc. Transl.* (2) **32** (1963), 37–81.

Landau, E.
[1] Abschätzung der Koeffizientensumme einer Potenzreihe. *Archiv Math. und Phys.* **21** (1913), 42–50; 250–255.
[2] "Darstellung und Begründung einiger neuerer Ergebnisse der Funktionentheorie" (Berlin, 1916). Reprinted by Chelsea, New York, 1946.
[3] Über die Blaschkesche Erweiterung des Vitalischen Satzes. *S.-B. Sächs. Akad. Wiss. Leipzig Math.-Natur. Kl.* **70** (1918), 156–159.
[4] Über einen Egerváryschen Satz. *Math. Z.* **29** (1929), 461.

Landsberg, M.
  [1] Lineare topologische Räume, die nicht lokalkonvex sind. *Math. Z.* **65** (1956), 104–112.

Lax, P. D.
  [1] Reciprocal extremal problems in function theory. *Comm. Pure Appl. Math.* **8** (1955), 437–453.

deLeeuw, K., and Rudin, W.
  [1] Extreme points and extremum problems in H_1. *Pacific J. Math.* **8** (1958), 467–485.

deLeeuw, K., Rudin, W., and Wermer, J.
  [1] The isometries of some function spaces. *Proc. Amer. Math. Soc.* **11** (1960), 694–698.

Littlewood, J. E.
  [1] On inequalities in the theory of functions. *Proc. London Math. Soc.* **23** (1925), 481–519.
  [2] On a theorem of Kolmogorov. *J. London Math. Soc.* **1** (1926), 229–231.
  [3] On a theorem of Zygmund. *J. London Math. Soc.* **4** (1929), 305–307.
  [4] Mathematical notes (13): On mean values of power series (II). *J. London Math. Soc.* **5** (1930), 179–182.
  [5] "Lectures on the Theory of Functions." Oxford Univ. Press, London and New York, 1944.

Livingston, A. E.
  [1] The space H^p, $0 < p < 1$, is not normable. *Pacific J. Math.* **3** (1953), 613–616.

Lohwater, A. J., Piranian, G., and Rudin, W.
  [1] The derivative of a schlicht function. *Math. Scand.* **3** (1955), 103–106.

Lohwater, A. J., and Ryan, F. B.
  [1] A distortion theorem for a class of conformal mappings. (To be published.)

Lumer, G.
  [1] "Algèbres de fonctions et espaces de Hardy." Lecture Notes in Mathematics, No. 75. Springer, Berlin, 1968.

Macintyre, A. J., and Rogosinski, W. W.
  [1] Some elementary inequalities in function theory. *Edinburgh Math. Notes* **35** (1945), 1–3.
  [2] Extremum problems in the theory of analytic functions. *Acta Math.* **82** (1950), 275–325.

Marden, M.
  [1] "Geometry of Polynomials." Amer. Math. Soc. Mathematical Surveys, No. 3; 2nd ed., 1966.

Nagasawa, M.
  [1] Isomorphisms between commutative Banach algebras with an application to rings of analytic functions. *Kōdai Math. Sem. Rep.* **11** (1959), 182–188.

Natanson, I. P.
  [1] "Theory of Functions of a Real Variable." Ungar, New York, 1955 and 1960.

Nehari, Z.
  [1] The Schwarzian derivative and schlicht functions. *Bull. Amer. Math. Soc.* **55** (1949), 545–551.
  [2] On bounded analytic functions. *Proc. Amer. Math. Soc.* **1** (1950), 268–275.
  [3] Extremal problems in the theory of bounded analytic functions. *Amer. J. Math.* **73** (1951), 78–106.
  [4] Bounded analytic functions. *Bull. Amer. Math. Soc.* **57** (1951), 354–366.
  [5] "Conformal Mapping." McGraw-Hill, New York, 1952.

Neuwirth, J., and Newman, D. J.
[1] Positive $H^{1/2}$ functions are constants. *Proc. Amer. Math. Soc.* **18** (1967), 958.

Nevanlinna, F. and R.
[1] Über die Eigenschaften analytischer Funktionen in der Umgebung einer singulären Stelle oder Linie. *Acta Soc. Sci. Fenn.* **50** (1922), No. 5.

Nevanlinna, R.
[1] Über beschränkte Funktionen die in gegebenen Punkten vorgeschrieben Werte annehmen. *Ann. Acad. Sci. Fenn.* **13** (1919), No. 1.
[2] Über beschränkte analytische Funktionen. *Ann. Acad. Sci. Fenn.* **32** (1929), No. 7.
[3] "Eindeutige Analytische Funktionen," 2nd ed. Springer, Berlin, 1953.

Newman, D. J.
[1] Interpolation in H^∞. *Trans. Amer. Math. Soc.* **92** (1959), 501–507.
[2] Some remarks on the maximal ideal structure of H^∞. *Ann. of Math.* **70** (1959), 438–445.
[3] The nonexistence of projections from L^1 to H^1. *Proc. Amer. Math. Soc.* **12** (1961), 98–99.
[4] Pseudo-uniform convexity in H^1. *Proc. Amer. Math. Soc.* **14** (1963), 676–679.

Newman, D. J. and Shapiro, H. S.
[1] The Taylor coefficients of inner functions. *Michigan Math. J.* **9** (1962), 249–255.

Noshiro, K.
[1] "Cluster Sets." Springer, Berlin, 1960.

Ostrowski, A.
[1] Über die Bedeutung der Jensenschen Formel für einige Fragen der komplexen Funktionentheorie. *Acta Sci. Math. (Szeged)* **1** (1922–23), 80–87.

Paley, R. E. A. C.
[1] On the lacunary coefficients of power series. *Ann. of Math.* **34** (1933), 615–616.

Paley, R. E. A. C., and Wiener, N.
[1] "Fourier Transforms in the Complex Domain." Amer. Math. Soc. Colloquium Publications, Vol. 19, 1934.

Paley, R. E. A. C., and Zygmund, A.
[1] On some series of functions, (1). *Proc. Cambridge Philos. Soc.* **26** (1930), 337–357.
[2] A note on analytic functions in the unit circle. *Proc. Cambridge Philos. Soc.* **28** (1932), 266–272.

Parreau, M.
[1] Sur les moyennes des fonctions harmoniques et analytiques et la classification des surfaces de Riemann. *Ann. Inst. Fourier (Grenoble)* **3** (1951), 103–197.

Penez, J.
[1] Approximation by boundary values of analytic functions. *Proc. Nat. Acad. Sci. U.S.A.* **40** (1954), 240–243.

Pick, G.
[1] Über die Beschränkungen analytischer Funktionen, welche durch vorgegebene Funktionswerte bewirkt werden. *Math. Ann.* **77** (1915), 7–23.
[2] Extremumfragen bei analytischen Funktionen im Einheitskreise. *Monatsh. Math.* **32** (1922), 204–218.

Piranian, G.
[1] Two monotonic, singular, uniformly almost smooth functions. *Duke Math. J.* 33 (1966), 255–262.

Plessner, A.
[1] Zur Theorie der konjugierten trigonometrischen Reihen. *Mitt. Math. Sem. Giessen* 10 (1923), 1–36.

Porcelli, P.
[1] "Linear Spaces of Analytic Functions." Rand McNally, Chicago, 1966.

Prawitz, H.
[1] Über Mittelwerte analytischer Funktionen. *Arkiv Mat. Astr. Fys.* 20 (1927–28), no. 6, 1–12.

Privalov, I. I.
[1] Intégrale de Cauchy. *Bulletin de l'Université, à Saratov,* 1918 (in Russian).
[2] Sur certaines propriétés métriques des fonctions analytiques. *Journal de l'École Polytechnique* 24 (1924), 77–112.
[3] Generalization of a theorem of Fatou. *Mat. Sb.* 31 (1924), 232–235 (in Russian).
[4] "Boundary Properties of Analytic Functions" (Moscow, 1941; 2nd ed., 1950). German transl.: Deutscher Verlag, Berlin, 1956.

Rado, T.
[1] "Subharmonic Functions." Ergebnisse der Mathematik und ihrer Grenzgebiete, Vol. 5 (1937). Reprinted by Chelsea, New York, 1949.

Riesz, F.
[1] Über Potenzreihen mit vorgeschriebenen Anfangsgliedern. *Acta Math.* 42 (1920), 145–171.
[2] Sur les valeurs moyennes du module des fonctions harmoniques et des fonctions analytiques. *Acta Sci. Math. (Szeged)* 1 (1922–23), 27–32.
[3] Über die Randwerte einer analytischen Funktion. *Math. Z.* 18 (1923), 87–95.
[4] Über eine Verallgemeinerung der Parsevalschen Formel. *Math. Z.* 18 (1923), 117–124.
[5] Sur une inégalité de M. Littlewood dans la théorie des fonctions. *Proc. London Math. Soc.* 23 (1925), 36–39.
[6] Sur la convergence en moyenne. *Acta Sci. Math. (Szeged)* 4 (1928–29), 58–64.
[7] Sur les valeurs moyennes des fonctions, *J. London Math. Soc.* 5 (1930), 120–121.
[8] Sur un théorème de maximum de MM. Hardy et Littlewood. *J. London Math. Soc.* 7 (1931), 10–13.

Riesz, F. and M.
[1] Über die Randwerte einer analytischen Funktion. *Quatrième Congrès des Math. Scand. Stockholm* (1916), 27–44.

Riesz, M.
[1] Sur les fonctions conjuguées. *Math. Z.* 27 (1927), 218–244.

Rogosinski, W. W.
[1] Über beschränkte Potenzreihen. *Compositio Math.* 5 (1937–38), 67–106; 442–476.

Rogosinski, W. W. and Shapiro, H. S.
[1] On certain extremum problems for analytic functions. *Acta Math.* 90 (1953), 287–318.

Romberg, B. W.
[1] The H_p spaces with $0 < p < 1$. Doctoral dissertation, Univ. of Rochester, 1960.

Royden, H. L.
[1] The boundary values of analytic and harmonic functions. *Math. Z.* **78** (1962), 1–24.
[2] "Real Analysis." Macmillan, New York, 1963.

Rudin, W.
[1] Analytic functions of class H_p. *In* "Lectures on Functions of a Complex Variable," pp. 387–397. Univ. of Michigan Press, Ann Arbor, 1955.
[2] Analytic functions of class H_p. *Trans. Amer. Math. Soc.* **78** (1955), 46–66.
[3] On a problem of Bloch and Nevanlinna. *Proc. Amer. Math. Soc.* **6** (1955), 202–204.
[4] Remarks on a theorem of Paley. *J. London Math. Soc.* **32** (1957), 307–311.
[5] The closed ideals in an algebra of analytic functions. *Canad. J. Math.* **9** (1957), 426–434.
[6] Trigonometric series with gaps. *J. Math. Mech.* **9** (1960), 203–227.
[7] "Principles of Mathematical Analysis," 2nd ed. McGraw-Hill, New York, 1964.
[8] "Real and Complex Analysis." McGraw-Hill, New York, 1966.
[9] "Function Theory in Polydiscs." Benjamin, New York, 1969.

Ryff, J. V.
[1] Subordinate H^p functions. *Duke Math. J.* **33** (1966), 347–354.

Sarason, D.
[1] The H^p spaces of an annulus. *Mem. Amer. Math. Soc.*, No. 56 (1965), 78 pp.

Schark, I. J.
[1] Maximal ideals in an algebra of bounded analytic functions. *J. Math. Mech.* **10** (1961), 735–746.

Schur, I.
[1] Über Potenzreihen, die im Innern des Einheitskreises beschränkt sind. *J. Reine Angew. Math.* **147** (1917), 205–232; **148** (1918), 122–145.

Shapiro, H. S.
[1] Applications of normed linear spaces to function-theoretic extremal problems. *In* "Lectures on Functions of a Complex Variable," pp. 399–404. Univ. of Michigan Press, Ann Arbor, 1955.
[2] Weakly invertible elements in certain function spaces, and generators in ℓ_1. *Michigan Math. J.* **11** (1964), 161–165.
[3] Remarks concerning domains of Smirnov type. *Michigan Math. J.* **13** (1966), 341–348.
[4] Monotonic singular functions of high smoothness. *Michigan Math. J.* **15** (1968), 265–275.

Shapiro, H. S., and Shields, A. L.
[1] On some interpolation problems for analytic functions. *Amer. J. Math.* **83** (1961), 513–532.

Shapiro, J.
[1] Linear functionals on non-locally convex spaces. Doctoral dissertation, Univ. of Michigan, 1969.

Smirnov, V. I.
[1] Sur la théorie des polynomes orthogonaux à une variable complexe. *Journal de la Société Phys.-Math. de Léningrade* **2** (1928), no. 1, 155–179.
[2] Sur les valeurs limites des fonctions, régulières à l'intérieur d'un cercle. *Journal de la Société Phys.-Math. de Léningrade* **2** (1929), no. 2, 22–37.
[3] Sur les formules de Cauchy et de Green et quelques problèmes qui s'y rattachent. *Izv. Akad. Nauk SSSR Ser. Mat.* 1932, 337–372.

[4] Über die Ränderzuordnung bei konformer Abbildung. *Math. Ann.* **107** (1932), 313–323.

Smirnov, V. I., and Lebedev, N. A.
[1] "Constructive Theory of Functions of a Complex Variable" (Moscow, 1964). English transl.: M.I.T. Press, Cambridge, Massachusetts, 1968.

Somadasa, H.
[1] Blaschke products with zero tangential limits. *J. London Math. Soc.* **41** (1966), 293–303.

Stein, E. M.
[1] Classes H^p, multiplicateurs et fonctions de Littlewood–Paley. *C. R. Acad. Sci. Paris* **263** (1966), 716–719; 780–781.

Stein, E. M., and Weiss, G.
[1] On the interpolation of analytic families of operators acting on H^p-spaces. *Tôhoku Math. J.* **9** (1957), 318–339.

Stein, E. M., and Zygmund, A.
[1] Boundedness of translation invariant operators on Hölder spaces and L^p-spaces *Ann. of Math.* **85** (1967), 337–349.

Stein, P.
[1] On a theorem of M. Riesz. *J. London Math. Soc.* **8** (1933), 242–247.

Sunouchi, G.
[1] Theorems on power series of the class H^p. *Tôhoku Math. J.* **8** (1956), 125–146.

Swinnerton-Dyer, H. P. F.
[1] On a conjecture of Hardy and Littlewood. *J. London Math. Soc.* **27** (1952), 16–21.

Szász, O.
[1] Ungleichungen für die Koeffizienten einer Potenzreihe. *Math. Z.* **1** (1918), 163–183.
[2] Ungleichheitsbeziehungen für die Ableitungen einer Potenzreihe, die eine im Einheitskreise beschränkte Funktion darstellt. *Math. Z.* **8** (1920), 303–309.
[3] On the coefficients of bounded power series. *Magyar Tudományos Akadémia Matematikai és Természettudományi Értesítöje* **43** (1926), 488–503 (in Hungarian; German summary).
[4] On bounded power series. *Magyar Tudományos Akadémia Matematikai és Természettudományi Értesítöje* **43** (1926), 504–520 (in Hungarian; German summary).

Szegö, G.
[1] Über die Randwerte einer analytischen Funktion. *Math. Ann.* **84** (1921), 232–244.
[2] Über orthogonale Polynome, die zu einer gegebenen Kurve der komplexen Ebene gehören. *Math. Z.* **9** (1921), 218–270.
[3] "Orthogonal Polynomials." Amer. Math. Soc. Colloquium Publications, Vol. 23, 2nd ed., 1959.

Sz.-Nagy, B., and Korányi, A.
[1] Relations d'un problème de Nevanlinna et Pick avec la théorie des opérateurs de l'espace hilbertien. *Acta Math. Acad. Sci. Hungar.* **7** (1956), 295–303.
[2] Operatortheoretische Behandlung und Verallgemeinerung eines Problemkreises in der komplexen Funktionentheorie. *Acta Math.* **100** (1958), 171–202.

Tanaka, C.
[1] On the class H_p of functions analytic in the unit circle. *Yokohama Math. J.* **4** (1956), 47–53.

Taylor, A. E.
[1] New proofs of some theorems of Hardy by Banach space methods. *Math. Mag.* **23** (1950), 115–124.
[2] Weak convergence in the spaces H^p. *Duke Math. J.* **17** (1950), 409–418.
[3] Banach spaces of functions analytic in the unit circle, I; II. Studia Math. 11 (1950), 145–170; 12 (1951), 25–50.

Taylor, G. D.
[1] A note on the growth of functions in H^p. *Illinois J. Math.* **12** (1968), 171–174.

Titchmarsh, E. C.
[1] "The Theory of Functions," 2nd ed. Oxford Univ. Press, London and New York, 1939.

Tumarkin, G. C.
[1] On integrals of Cauchy–Stieltjes type. *Uspehi Mat. Nauk.* **11** (1956), no. 4 (70), 163–166 (in Russian).
[2] Approximation in the mean to functions on rectifiable curves. *Mat. Sb.* **42** (84) (1957), 79–128 (in Russian).
[3] A sufficient condition for a domain to belong to class S. *Vestnik Leningrad. Univ.* **17** (1962), no. 13, 47–55 (in Russian).

Tumarkin, G. C., and Havinson, S. Ja.
[1] On the removal of singularities for analytic functions of a certain class (class D). *Uspehi Mat. Nauk* 12 (1957), no. 4 (76), 193–199 (in Russian).
[2] On the definition of analytic functions of class E_p in multiply connected domains. *Uspehi Mat. Nauk* 13 (1958), no. 1 (79), 201–206 (in Russian).
[3] Classes of analytic functions in multiply connected domains represented by the formulas of Cauchy and Green. *Uspehi Mat. Nauk* 13 (1958), no. 2 (80), 215–221 (in Russian).
[4] On the decomposition theorem for analytic functions of class E_p in multiply connected domains. *Uspehi Mat. Nauk* 13 (1958), no. 2 (80), 223–228 (in Russian).
[5] Analytic functions in multiply connected domains of Smirnov type (class S). *Izv. Akad. Nauk SSSR Ser. Mat.* **22** (1958), 379–386 (in Russian).
[6] Investigation of the properties of extremal functions by means of duality relations in extremal problems for classes of analytic functions in multiply connected domains. *Mat. Sb.* **46** (88) (1958), 195–228 (in Russian).
[7] Classes of analytic functions in multiply connected domains. *In* "Issledovaniya po sovremennym problemam teorii funkciĭ kompleksnogo peremennogo" (Moscow, 1960), 45–77 (in Russian). French transl.: *In* "Fonctions d'une variable complexe. Problèmes contemporains," pp. 37–71. Gauthiers-Villars, Paris, 1962.
[8] Qualitative properties of solutions to certain types of extremal problems. *In* "Issledovaniya po sovremennym problemam teorii funkciĭ kompleksnogo peremennogo" (Moscow, 1960), 77–95 (in Russian). French transl.: *In* "Fonctions d'une variable complexe. Problèmes contemporains," pp. 73–92. Gauthiers-Villars, Paris, 1962.

Voichick, M.
[1] Ideals and invariant subspaces of analytic functions. *Trans. Amer. Math. Soc.* **111** (1964), 493–512.
[2] Invariant subspaces on Riemann surfaces. *Canad. J. Math.* **18** (1966), 399–403.
[3] Extreme points of bounded analytic functions on infinitely connected regions. *Proc. Amer. Math. Soc.* **17** (1966), 1366–1369.

Voichick, M., and Zalcman, L.
[1] Inner and outer functions on Riemann surfaces. *Proc. Amer. Math. Soc.* **16** (1965), 1200–1204.

Walters, S. S.
[1] The space H^p with $0 < p < 1$. *Proc. Amer. Math. Soc.* **1** (1950), 800–805.
[2] Remarks on the space H^p. *Pacific J. Math.* **1** (1951), 455–471.

Weiss, G.
[1] An interpolation theorem for sublinear operators on H_p spaces. *Proc. Amer. Math. Soc.* **8** (1957), 92–99.

Weiss, M. and G.
[1] A derivation of the main results of the theory of H^p spaces. *Rev. Un. Mat. Argentina* **20** (1962), 63–71.

Wells, J. H.
[1] Some results concerning multipliers of H^p. *J. London Math. Soc.* **2** (1970).

Wermer, J.
[1] Banach algebras and analytic functions. *Advances in Math.* **1** (1961), fasc. 1, 51–102.

Widder, D. V.
[1] "The Laplace Transform." Princeton Univ. Press, Princeton, New Jersey, 1941.

Zygmund, A.
[1] Sur les fonctions conjuguées. *Fund. Math.* **13** (1929), 284–303; corr. **18** (1932), 312.
[2] Smooth functions. *Duke Math. J.* **12** (1945), 47–76.
[3] On the preservation of classes of functions. *J. Math. Mech.* **8** (1959), 889–895.
[4] "Trigonometric Series." Cambridge Univ. Press, London and New York, 1959.

AUTHOR INDEX

SUBJECT INDEX

Pure and Applied Mathematics

A Series of Monographs and Textbooks

Editors

Paul A. Smith and Samuel Eilenberg

Columbia University, New York

Pure and Applied Mathematics

A Series of Monographs and Textbooks

In preparation

EDUARD PRUGOVECKI. Quantum Mechanics in Hilbert Space.

BODO PAREIGIS. Categories and Functors: An Introduction